Medical Intelligence Unit

Cell Adhesion and Invasion in Cancer Metastasis

Pnina Brodt, Ph.D.

Department of Surgery
Division of Surgical Research
McGill University
Montreal, Quebec, Canada

CHAPMAN & HALL
I(T)P An International Thomson Publishing Company

New York • Albany • Bonn • Boston • Cincinnati • Detroit • London • Madrid • Melbourne •
Mexico City • Pacific Grove • Paris • San Francisco • Singapore • Tokyo • Toronto • Washington

R.G. Landes Company
Austin

Medical Intelligence Unit

CELL ADHESION AND INVASION IN CANCER METASTASIS

R.G. LANDES COMPANY
Austin, Texas, U.S.A.

Printed in the U.S.A.

Please address all inquiries to the Publishers:
R.G. Landes Company, 909 Pine Street, Georgetown, Texas, U.S.A. 78626
Phone: 512/ 863 7762; FAX: 512/ 863 0081

North American distribution:
Chapman & Hall, 115 Fifth Avenue, New York, New York, U.S.A. 10003

International distribution (except North America):
Springer-Verlag GmbH & Co. KG, Tiergartenstrasse 17, D-69121 Heidelberg, Germany

U.S. and Canada ISBN: 0-412-10151-3
International ISBN: 3-540-60686-6

Library of Congress Cataloging-in-Publication Data

Cell adhesion and invasion mechanisms in cancer metastasis / [edited by] Pnina Brodt.
p. cm. — (Medical intelligence unit)
Includes bibliographical references and index.
ISBN 0-412-10151-3 (alk. paper)
1. Cancer invasiveness. 2. Metastasis. 3. Cell adhesion. 4. Cell adhesion molecules.
5. Cancer cells. I. Brodt, Pnina, 1950- . II. Series.
[DNLM: 1. Neoplasm Metastasis--physiopathology. 2.Cell Adhesion.
3. Neoplasm Invasiveness. QZ 202 C3923 1996]
RC268.5.C386 1996
616.99'407—dc20
DNLM/DLC
for Library of Congress

95-46850
CIP

Publisher's Note

R.G. Landes Company publishes six book series: *Medical Intelligence Unit, Molecular Biology Intelligence Unit, Neuroscience Intelligence Unit, Tissue Engineering Intelligence Unit, Environmental Intelligence Unit* and *Biotechnology Intelligence Unit.* The authors of our books are acknowledged leaders in their fields and the topics are unique. Almost without exception, no other similar books exist on these topics.

Our goal is to publish books in important and rapidly changing areas of bioscience for sophisticated researchers and clinicians. To achieve this goal, we have accelerated our publishing program to conform to the fast pace in which information grows in bioscience. Most of our books are published within 90 to 120 days of receipt of the manuscript. We would like to thank our readers for their continuing interest and welcome any comments or suggestions they may have for future books.

Deborah Muir Molsberry
Publications Director
R.G. Landes Company

CONTENTS

EDITOR

Pnina Brodt, Ph.D.
Department of Surgery
Division of Surgical Research
McGill University
Montreal, Quebec, Canada
Chapters 1, 4, 7, 9, 11

CONTRIBUTORS

Orest W. Blaschuk
Division of Urology
Department of Surgery
McGill University
Royal Victoria Hospital
Montreal, Quebec, Canada
Chapter 3

Shoukat Dedhar
Division of Cancer Biology and Department of Medical Biophysics
University of Toronto
Sunnybrook Health Science Center
Toronto, Ontario, Canada
Chapter 4

Margaret Durko
Department of Surgery
McGill University
Royal Victoria Hospital
Montreal, Quebec, Canada
Chapter 9

Tanya N. Mayadas
Vascular Research Division
Department of Pathology
Brigham and Women's Hospital and Harvard Medical School
Boston, Massachusetts, U.S.A.
Chapter 6

John S. Mort
Joint Diseases Laboratory
Shriners Hospital for Crippled Children and Department of Surgery
McGill University
Montreal, Quebec, Canada
Chapter 10

Sandra B. Munro
Division of Urology
Department of Surgery
McGill University
Royal Victoria Hospital
Montreal, Quebec, Canada
Chapter 3

John Nip
Division of Surgical Research
Department of Pathology
McGill University
Montreal, Quebec, Canada
Chapter 7

Shafaat A. Rabbani
McGill University
Royal Victoria Hospital
Montreal, Quebec, Canada
Chapter 8

CONTRIBUTORS

Anneliese D. Recklies
Joint Diseases Laboratory
Shriners Hospital for Crippled Children
and Department of Surgery
McGill University
Montreal, Quebec, Canada
Chapter 10

Chi-Hung Siu
Banting and Best Department of Medical Research
University of Toronto
Toronto, Ontario, Canada
Chapters 2, 5

PREFACE

Approximately 20 years ago tumor biologists began to realize that different tumor cells had different patterns of cell adhesion. Some adhered well to plastic or glass dishes while others did not. Some adhered tightly to other tumor cells of the same type forming large multicellular aggregates that floated in tissue culture medium. Still others adhered well to neither themselves, nor to plastic or glass but could adhere to different cell types such as hepatocytes, brain cells or vascular endothelial cells. These early results were somewhat confusing and it was unclear whether cell adhesion was positively or negatively correlated with tumor invasion and metastasis. It is now clear that adhesive events are an integral component of the metastatic phenotype and that certain types of cell adhesion promote metastasis, whereas other types inhibit the metastatic process. In general, increased homotypic adhesion of tumor cells to other tumor cells in the primary tumor site appears to prevent cellular exit from the primary tumor and thus, inhibits metastasis. In contrast, tumor cell adhesion to endothelial cells or subendothelial matrix components represent important early steps of the extravasation process and consequently, serve to promote metastasis.

The critical work in this field in the past decade has come in two waves. The first was the establishment of correlations between cell adhesion and tumor invasion/metastasis. Such studies have shown that not only is tumor adhesion to the vasculature an essential component of metastasis, it also has a critical role in the preferential establishment of particular tumor metastases in specific organ sites. The second important set of findings has to do with the nature of the particular molecules that are involved in tumor cell adhesion. These include the cadherins that mediate cell-cell interactions, the integrins that mediate cell-cell and cell-extracellular matrix interactions, cell adhesion receptors of the immunoglobulin superfamily that have a role in tumor cell adhesion to the endothelium, and selectins that can be preferentially upregulated on specific endothelial cell beds.

In the current volume, Dr. Brodt and her colleagues do a superb job of providing an up-to-date critical review of the latest information on the specific cell adhesion molecules involved in the metastatic process. Advances in this field have been extremely rapid and this volume serves as an excellent compendium of the more important recent findings regarding these adhesion molecules. The introductory chapters by Drs. Brodt and Siu give an admirable summary of the important questions and concepts in this field with detailed chapters following on the molecular biology and biochemistry of the cadherins by Drs. Munro and Blaschuk, the integrins by Drs. Dedhar and Brodt, the selectins and their ligands by Mayadas, and the Ig-like adhesion family by Dr. Siu. In addition, the later chapters provide both methodological insight and conceptual overview of the role of proteases and protease inhibitors in regulating the metastatic process and modulating cell adhesion. The excellent chapters by Drs. Rabbani, Durko, Brodt and Mort, provide a comprehensive examination of the evidence for a role for urokinase-plasminogen activator, metalloproteinases, and lysosomal proteases in regulating cell invasion, and Dr. Nip's chapter provides a concise overview of the current methods available for the analysis of cell invasion. In all, this new volume provides both an excellent introduction along with an up-to-date summary of the current research in cell adhesion and tumor-mediated matrix degradation in facilitating tumor invasion and metastasis. The book contains useful information for both the novice and the expert and certainly represents the most comprehensive and insightful discussion of these topics that is available today.

Bruce R. Zetter
Department of Surgery
and Physiology
Harvard Medical School
and Children's Hospital
Boston, Massachusetts, U.S.A.

BIOGRAPHICAL SKETCH

Dr. Pnina Brodt is an associate professor in the Departments of Surgery and Oncology at McGill University and a former Medical Research Council of Canada and Fonds de la Recherche en Santé du Quebec Scholar. Dr. Brodt earned her B.Sc. in Microbiology and Biochemistry at Bar-Ilan University in Israel. She earned her M.Sc. in Microbiology at the University of Ottawa and a Ph.D. in Microbiology and Immunology at McGill University. She has been a Research Fellow of the Medical Research Council of Canada at the Weizmann Institute of Science in Israel and at McGill University and a Visiting Scholar at the Dana Farber Cancer Institute and Harvard Medical School. She is the author of numerous original papers and several reviews in Tumor Immunology and Cancer Biology. She is currently involved in research on the role of cell adhesion and autocrine and paracrine growth regulation in the process of metastasis and lectures on the Biology of Cancer in several graduate and undergraduate courses.

ACKNOWLEDGMENTS

This book was made possible through the cooperation of several individuals. I am indebted to my colleagues for contributing excellent and timely reviews to this collection and for meeting production deadlines despite their heavy schedules. I am grateful to John Nip for his untiring and invaluable assistance in the review of the literature and for the excellent illustrations and to Mrs. Irene Sidorenko for her assistance in the preparation of manuscripts. Many thanks also go to my husband Abraham and my children Eliezer, Shoshana, Chaim and Nilly for their understanding, encouragement and support (and for staying out of my way long enough to finish this work). To all of you, I hope this book will be, in some measure, a fitting reward for your efforts.

PART I

MOLECULAR MEDIATORS OF CELL ADHESION AND EXTRACELLULAR MATRIX PROTEOLYSIS: STRUCTURE, FUNCTION AND REGULATION

CHAPTER 1

General Introduction

Pnina Brodt

Cellular communication, i.e. the ability of individual cells to transmit and receive signals from their microenvironment, is the key to many of the biological processes required for normal development and homeostasis. It is an essential component of embryogenesis, inflammation, wound healing, blot clotting and the host immune response. These signals may be transmitted via diffusible factors which bind to appropriate receptors on the cell surface. Often, however, they depend on direct contact between the cell and its partner in the signaling process which may be either another cell or a component of the extracellular matrix (reviewed in refs. 1-5).

Cell-cell and cell-matrix interactions are tightly regulated processes which can in turn regulate cell shape, determine cell polarity and tissue architecture, trigger cell motility, and induce the expression of receptors for growth factors thereby affecting cellular proliferation. Rapid progress in the past decade in the study of these receptors revealed that, despite the wide spectrum of physiological mechanisms which require cellular adhesion and the diverse types of cells involved, the adhesion molecules are remarkably similar. Thus, early events in the development of the nervous system[6,7] are driven by the same molecules or members of the same families of molecules which mediate graft rejection[8] while adhesion molecules, which play a role in the development of the gut epithelium,[9] are similar to endothelial cell receptors mediating leukocyte migration.[10]

Cell Adhesion and Invasion in Cancer Metastasis, edited by Pnina Brodt.

The majority of these adhesion receptors have now been grouped into four main families on the basis of sequence homology and structural similarity. These include the cadherins[11] (see chapter 3), integrins[12] (see chapter 4), members of the Ig superfamily[13] (see chapter 5) and selectins[14] (see chapter 6). Individual members of these families may serve as adhesion ligands for members of the same or other families.[5,10,15,16] They may act coordinately to mediate the same process[10,17-19] or they may trigger events which regulate the expression and function of other receptors and cellular mediators such as growth factor receptors and proteolytic enzymes.[20-22] Often, therefore, it is the sum of the functions of multiple adhesion receptors which will determine the final outcome of a cellular adhesion event, and it is only through the integrated study of these diverse functions that a clearer understanding of the molecular control and consequences of cell adhesion will eventually emerge.

One area in which the recent developments in the study of cell adhesion mechanisms have had a major beneficial impact is oncology. It has become clear that the process of malignant progression is accompanied from its onset by profound changes in the expression and function of cell adhesion molecules.[23] Malignant transformation may trigger the expression of new receptors,[24] result in the loss or suppression of others,[25] or lead to expression of genetic variants of receptors expressed on normal cells.[26] These changes eventually give rise to cells with modified repertoires of adhesion receptors whose potential to communicate with the microenvironment has been irreversibly altered.[23,27] These alterations are at the root of the cancer cell's potential to dissociate from its neighboring cells, migrate through the stroma, invade and exit blood vessels, anchor in a foreign microenviroment and respond to growth modulating cytokines in a modified manner—processes which are the hallmark of cancer progression and metastasis.

In view of the central role which cell adhesion plays in the process of cancer dissemination, it is of little surprise that studies on the involvement of cell adhesion in malignancy are presently pursued in many laboratories. In different centers attempts are in progress to utilize the rapidly growing body of information in this area towards the development of new reagents and strategies for diagnosis, prognosis and clinical intervention in the process of cancer metastasis.[28-31]

The aim of this book is to summarize recent developments in this field with emphasis on the regulatory role which cell adhesion mechanisms play in tumor cell migration and invasion and in the regulation of tumor cell growth—processes which underlie cancer metastasis. While descriptive studies dealing with specific changes in the expression of adhesion receptors on different malignancies are undoubtedly of great importance to the overall understanding of malignant progression they will be summarized here only briefly and be dealt with at length only to the extent that they can shed light on the functional significance of an adhesion receptor or family of receptors in invasion and metastasis.

The book has been arranged into two major parts. The first part consists of short reviews each providing a summary of the current state of knowledge on a major family of adhesion receptors or proteinases implicated in the pathology of malignant disease. These reviews dealing with structure/function aspects and the regulation of these mediators provide the background and basis for the second part which more specifically reviews the involvement of these molecules in cancer progression, angiogenesis, invasion and metastasis. A comprehensive, all-inclusive review of this area of research is beyond the scope of this work. Omissions wherever they occurred in no way reflect a value judgment by the authors but are rather indicative of the enormous body of excellent literature published in the field, particularly in the past two decades. In order to avoid duplication of other works, an attempt has been made to focus here on newly emerging trends in our view of the role of adhesion receptors in malignancy. These include the evidence for the existence of adhesion cascades and regulatory networks involving adhesion receptors, proteolytic enzyme growth factors and their receptors.

References

1. Damsky C, Sutherland A, Fisher S. Extracellular matrix-5—adhesive interactions in early mammalian embryogenesis, implantation, and placentation. FASEB J 1993; 7:1320-29.
2. Glukhova MA, Thiery JP. Fibronectin and integrins in development. Sem Cancer Biol 1993; 4:241-49.
3. Sligh JE, Ballantyne CM, Rich SS et al. Inflammatory and immune responses are impaired in mice deficient in intercellular adhesion molecule-1. PNAS USA 1993; 90:8529-33.

4. Ginsberg MH, Du XP, O' Toole TE et al. Platelet integrins. Throm Haem 1993; 70:87-93.
5. Mackay CR, Imhof BA. Cell adhesion in the immune system. Immunol Today 1993; 14:99-102.
6. Venstrom KA, Reichardt LF. Extracellular matrix .2: role of extracellular matrix molecules and their receptors in the nervous system. FASEB J 1993; 7:996-1003.
7. Nybroe O, Bock E. Structure and function of the neural cell adhesion molecules NCAM and L1. Adv Exp Med Biol 1990; 265:185-96.
8. Lin Y, Kirby JA, Browell DA et al. Renal allograft rejection—expression and function of VCAM-1 on tubular epithelial cells. Clin Exp Immunol 1993; 92:145-51.
9. Hunkapiller TH, Hood L. Diversity of the immunoglobulin gene superfamily. Adv Immunol 1989; 44:1-63.
10. Bevilacqua MP. Endothelial-leukocyte adhesion molecules. Ann Rev Immunol 1993; 11:767-804.
11. Takeichi M. Cadherin cell adhesion receptors as a morphogenetic regulator. Science 1991; 251:1451-55.
12. Hynes RO. Integrins: versatility, modulation, and signaling in cell adhesion. Cell 1992; 69:11-25.
13. Buck CA. Immunoglobulin superfamily: structure, function and relationship to other receptor molecules. Sem Cell Biol 1992; 3:179-88.
14. Bevilacqua MP, Nelson RM. Selectins. J Clin Invest 1993; 91:379-387.
15. Campanero MR, Delpozo MA, Arroyo AG et al. ICAM-3 interacts with LFA-1 and regulates the LFA-1/ICAM-1 cell adhesion pathway. J Cell Biol 1993; 123:1007-16.
16. Chan PY, Aruffo A. VLA-4 integrin mediates lymphocyte migration on the inducible endothelial cell ligand VCAM-1 and the extracellular matrix ligand fibronectin. J Biol Chem 1993; 268:24655-64.
17. Bauer JS, Schreiner CL, Giancotti FG et al. Motility of fibronectin receptor-deficient cells on fibronectin and vitronectin: collaborative interactions among integrins. J Cell Biol 1992; 116:477-87.
18. Hahne M, Jager U, Isenmann S et al. Five tumor necrosis factor-inducible cell adhesion mechanisms on the surface of mouse endothelioma cells mediate the binding of leukocytes. J Cell Biol 1993; 121:655-64.
19. Picker LF, Treer JF, Nguyen M et al. Coordinate expression of β_1 and β_2 integrin activation epitopes during T cell responses in secondary lymphoid tissue. Eur J Immunol 1993; 23:2751-57.
20. Werb Z, Tremble PM, Behrendtsen O et al. Signal transduction through the fibronectin receptor induces collagenase and stromelysin gene expression. J Cell Biol 1989; 109:877-89.

21. Seftor REB, Seftor EA, Stetler-Stevenson WG et al. The 72 kDa Type IV collagenase is modulated via differential expression of $\alpha_5\beta_1$ integrins during human melanoma cell invasion. Cancer Res 1993; 53:3411-5.
22. Schwartz MA, Ingber DE. Integrating with integrins. Mol Biol of the Cell 1994; 5:389-93.
23. Albelda SM. Biology of disease. Role of integrins and other cell adhesion molecules in tumor progression and metastasis. Lab Invest 1993; 68:4-17.
24. Albelda SM, Mette SA, Elder DE et al. Integrin distribution in malignant melanoma: association of the β_3 subunit with tumor progression. Cancer Res 1990; 50:6757-64.
25. Birchmeier W, Behrens J, Weidner KM et al. Dominant and recessive genes involved in tumor cell invasion. Curr Opin Cell Biol 1991; 3:832-40.
26. Herrlich P, Zoller M, Pals ST et al. CD44 splice variants—metastases meet lymphocytes. Immunol Today 1993; 14:395-399.
27. Zetter BR. Adhesion molecules in tumor metastasis. Sem Cancer Biol 1993; 4:219-29.
28. Pignatelli M, Vessey CJ. Adhesion molecules: novel molecular tools in tumor pathology. Human Path 1994; 25:849-56.
29. Humphries MJ. Fibronectin and cancer—rationales for the use of anti adhesives in cancer treatment. Sem Cancer Biol 1993; 4:293-9.
30. Weinstat-Saslow D, Steeg PS. Angiogenesis and colonization in the tumor metastatic process: basic and applied advances. FASEB J 1994; 8:401-7.
31. Goldfarb RH, Brunson KW. Therapeutic agents for treatment of established metastases and inhibitors of metastatic spread: preclinical and clinical progress. Curr Opin Oncol 1992; 4:1130-41.

CHAPTER 2

Cell Adhesion Receptors: A General Introduction

Chi-Hung Siu

INTRODUCTION

Cellular adhesive property and substrate environment are crucial to morphogenesis and organogenesis from the very early stages of embryonic development. Altered adhesive properties often lead to disease states. Cell-cell adhesion and cell-substrate adhesion are mediated by two major categories of molecules, the cell adhesion molecules (CAMs) and the cell substrate adhesion molecules (SAMs), respectively. The interplay of CAMs and SAMs is vital to a large number of normal and pathological phenomena, such as neurite outgrowth and targeting, lymphocyte homing and infiltration, bone repair, tissue regeneration, and metastasis of tumor cells.[1-7] In addition to their contribution to the cell-cell and cell-substrate interactions, there is growing evidence that both CAMs and SAMs are involved in signaling processes in a variety of cell types.[7-9]

Since the pioneering work of Gerisch and coworkers in the use of stage-specific antibodies to define cell adhesion sites in *Dictyostelium,*[10] with the advent of monoclonal antibody technology, the use of antibodies to perturb cell adhesion has proven to be immensely successful in the identification and purification of numerous membrane and matrix components involved in cell-cell and cell-substrate interactions. These antibodies have provided some powerful tools for the immunolocalization of CAMs and SAMs,

Cell Adhesion and Invasion in Cancer Metastasis, edited by Pnina Brodt.

the isolation of adhesion complexes by immunoprecipitation, and the molecular cloning of CAM and SAM cDNAs from expression libraries.

MAJOR CLASSES OF ADHESION RECEPTORS ON THE CELL SURFACE

There are four major classes of adhesion receptors on the cell surface: (1) integrins, (2) selectins, (3) cadherins, and (4) members of the immunoglobulin (Ig) superfamily. All integrins are composed of noncovalently associated αβ heterodimers, that are known to mediate leukocyte-leukocyte and leukocyte-endothelial cell adhesion,[11,12] as well as cellular interactions with extracellular matrix components, such as fibronectin, laminin, collagen, and fibrinogen.[3,7] The selectins are lectin-like molecules that bind carbohydrates in a Ca^{2+}-dependent manner.[12,13] The cadherin family is a group of adhesion receptors that mediates calcium-dependent cell-cell adhesion via homophilic interaction.[6,14] The Ig superfamily includes a large number of related molecules involved in cell-cell adhesion.[15,16] They all share a common Ig homology unit, which is characterized by an amino acid sequence ~100 amino acids in length and organized into the so-called antibody fold. Members of the Ig superfamily participate in a variety of homophilic and heterophilic interactions via their polypeptide domains.[15]

ASSAYS FOR THE STUDY OF CELL ADHESION MECHANISMS

Although the inhibition of cell-cell adhesion or cell-substrate adhesion by specific antibodies has led to the identification of many SAMs and CAMs, this approach does not distinguish between specific binding and nonspecific steric hindrance. Therefore, it is necessary to provide direct demonstration of the cell binding property for these molecules. Assays that are used to investigate the cell binding activity of SAMs and CAMs can be further exploited to elucidate the molecular mechanisms involved in cell adhesion. For instance, most CAMs in the Ig superfamily are known to mediate cell-cell adhesion via Ca^{2+}-independent homophilic interactions. However, several Ig CAMs have been shown to interact heterophilically with members of the integrin family.[17,18] Consensus glycosaminoglycan recognition sequences[19] are found in several Ig CAMs, suggesting that their interactions with proteoglycans may also contribute to the adhesiveness between cells. It is, therefore, important to distinguish between these different mechanisms and to assess their contribution to intercellular adhesion. A brief summary of the more commonly used assays and strategies to investigate CAM action is given below. Most of these assays are also applicable to SAMs.

First, a purified CAM can be radiolabeled for determination of its specific binding to cell surface receptors.[20,21] The binding should be dose-dependent and saturable. Usually, competition experiments are included to demonstrate specificity. The requirement of metal cations can also be assessed.[22] To distinguish between homophilic binding and heterophilic binding, cells can be precoated with anti-CAM antibodies before the addition of labeled CAM for binding. Inhibition of binding is consistent with a homophilic binding mechanism, while an unaltered level of binding implies heterophilic interaction.[20]

Second, the conjugation of purified CAM to Covaspheres provides a simple method to visualize CAM-mediated binding activity to a specific cell type.[20,23,24] Binding of these microspheres can be monitored easily by fluorescence microscopy and quantified by fluorescence-activated cell sorting. Alternatively, purified CAM can be incorporated into lipid vesicles for attachment assays.[25] The effects of CAM concentration on the rate of vesicle aggregation will also provide an important indicator of its adhesive property.[26]

Third, the effects of a CAM on cell attachment and spreading can be investigated by a cell-to-substratum attachment assay. The CAM can be covalently conjugated to a substratum[27] or adsorbed to a nitrocellulose-coated surface.[28] The attachment of either fluorescein-labeled or radioactively labeled cells can be scored microscopically or by scintillation counting. Alternatively, cells can be deposited for attachment onto a monolayer of cells expressing a specific CAM.[29,30] Competition experiments can be carried out using specific antibodies or peptides.

Fourth, the centrifugal force-based adhesion assay has been used to assess the relative adhesiveness of cells to immobilized CAM.[31,32] Purified CAM is used to coat the bottom of microtiter plate wells. Cells labeled metabolically are added and allowed to attach to these wells. Another set of wells, filled with medium, is inverted on the first. The assembly is sealed, inverted and subjected to centrifugation. The relative number of cells retained in the first well will be estimated and the centrifugal force required to shear off the attached cells can be used to calculate the adhesive strength of a given CAM.

Fifth, transfection of cells which normally do not express CAMs has provided a powerful approach for the analysis of CAM function.[33-41] CAMs expressed in mouse L cells or NIH 3T3 cells are able to function in the natural environment of a plasma membrane, leading to the formation of cell aggregates in suspension cultures. Whether a given CAM binds heterophilically with a receptor or homophilically with the same CAM on apposing cells can be readily distinguished. Subtle differences in binding affinity between two closely related CAMs can be investigated by the sorting out of cells in coaggregates composed of two separately transfected cell populations.[42] Transfected cells can be cultured as a monolayer to provide a substratum for the attachment of CAM-vesicles, CAM-conjugated Covaspheres or specific cells. These assays have also been applied in the study of inductive events elicited by CAMs in several coculture systems.[43,44]

Several strategies have been employed successfully in the identification of cell adhesion sites in CAMs. First, mapping of epitopes of monoclonal antibodies that inhibit the cell binding activity of a given CAM can lead to the mapping of functional domains involved in homophilic or heterophilic interactions.[45-48] This can be achieved by identifying the smallest proteolytic fragment that harbors the epitope. If the antibody recognizes a linear epitope, it can be mapped by using short overlapping peptides assembled on plastic pins in solid-state ELISA.[49] Next, fusion proteins or synthetic oligopeptides covering the region of interest in binding and/or competition experiments have been used to narrow down the peptide sequence participating directly in cell binding.[48,50] Then peptide analogues can be used to investigate structural characteristics and side-chain requirements in the binding reaction. The peptide approach is most useful if the binding site consists of only a short stretch of residues. However, results may not be easily interpretable, if the binding site is made up of peptide sequences from several distantly separated regions. Finally, site-directed mutagenesis can be used to confirm results obtained by the above approaches and to define new secondary sites that may act synergistically with the principal binding site.[51-54] The limitation here is that even single mutations may sometimes give rise to substantial changes in the three-dimensional structure of the protein; thus the loss of function may not be a direct result of the mutation. It is, therefore, evident that a combination of several approaches needs to be taken before a valid conclusion can be drawn. Ultimately, a more precise picture of these molecular interactions will depend on the use of high resolution structural analyses, such as nuclear magnetic resonance and X-ray crystallography.[55-58]

References

1. Albelda SM. Endothelial and epithelial cell adhesion molecules. Am J Respir Cell Mol Biol 1991; 4:195-203.
2. Edelman GM, Crossin KA. Cell adhesion molecules: implications for a molecular histology. Ann Rev Biochem 1991; 60:155-90.
3. Hynes RO. Integrins: a family of cell receptors. Cell 1987; 48:549-54.
4. Hynes RO, Lander AD. Contact and adhesive specificities in the associations, migrations, and targeting of cells and axons. Cell 1992; 68:303-22.
5. Singer SJ. Intercellular communication and cell-cell adhesion. Science 1992; 255:1671-77.
6. Takeichi M. Cadherins: a molecular family important in selective cell-cell adhesion. Ann Rev Biochem 1990; 59:237-52.
7. Hynes RO. Integrins: versatility, modulation, and signaling in cell adhesion. Cell 1992; 69:11-25.
8. Doherty P, Walsh FS. CAMs, second messengers, and axonal growth. Curr Opin Neurobiology 1992; 2:595-601.
9. Walsh FS, Doherty P. Factors regulating the expression and function of calcium-independent cell adhesion molecules. Curr Opin Cell Biol 1993: 5:791-96.
10. Beug H, Katz F, Gerisch G. Dynamics of antigenic membrane sites relating to cell aggregation in *Dictyostelium*. J Cell Biol 1973; 56:647-58.
11. Osborn L. Leukocyte adhesion to endothelium in inflammation. Cell 1990; 62:3-6.
12. McEver, RP. Leukocyte-endothelial cell interactions. Curr Opin Cell Biol 1992; 4:840-9.
13. Brandley BK, Swiedler SJ, Robbins PW. Carbohydrate ligands of the LEC cell adhesion molecules. Cell 1990; 63:861-3.
14. Takeichi M. Cadherin cell adhesion receptors as a morphogenetic regulator. Science 1991; 251:1451-5.
15. Williams AF, Barclay AN. The immunoglobulin superfamily: domains for cell surface recognition. Annu Rev Immunol 1988; 6: 381-405.
16. Hunkapiller T, Hood L. Diversity of the immunoglobulin gene superfamily. Adv Immunol 1989; 44:1-63.

17. Elices MJ, Osborn L, Takada Y et al. VCAM-1 on activated endothelium interacts with the leukocyte integrin VLA-4 at a site distinct from the VLA4/fibronectin binding site. Cell 1990; 60:577-84.
18. Springer TA. Adhesion receptors of the immune system. Nature 1990; 346:425-33.
19. Cardin AD, Weintraub HJR. Molecular modeling of protein of protein-glycosaminoglycan interactions. Arteriosclerosis 1988; 9:21-32.
20. Siu CH, Cho A, Choi AHC. The contact site A glycoprotein mediates cell-cell adhesion by homophilic binding in *Dictyostelium discoideum*. J Cell Biol 1987; 105:2523-33.
21. Gao EN, Shier P, Siu CH. Purification and partial characterization of a cell adhesion molecule (gp150) involved in postaggregation stage cell-cell binding in *Dictyostelium discoideum*. J Biol Chem 1992; 267:9409-15.
22. Brar SK, Siu CH. Characterization of the cell adhesion molecule gp24 in *Dictyostelium discoideum*: mediation of cell-cell adhesion via a Ca^{2+}-dependent mechanism. J Biol Chem 1993; 268:24902-9.
23. Rutishauser U, Hoffman S, Edelman GM. Binding properties of a cell adhesion molecule from neural tissue. Proc Natl Acad Sci USA 1982; 79:685-9.
24. Grumet M, Edelman GM. Neuron-glia cell adhesion molecule interacts with neurons and astroglia via different binding mechanisms. J Cell Biol 1988; 106:487-503.
25. Johnson PW, Abramow-Newerly W, Seilheimer MM et al. Recombinant myelin-associated glycoprotein confers neural adhesion and neurite outgrowth function. Neuron 1989; 3:377-85.
26. Hoffman S, Edelman GM. Kinetics of homphilic binding by embryonic and adult forms of the neural cell adhesion molecule. Proc Natl Acad Sci USA 1983; 80:5762-6.
27. Cole GJ, Schubert D, Glaser L. Cell-substratum adhesion in chick neural retinal cells depends upon protein-heparan sulfate interactions. J Cell Biol 1985; 100:1192-9.
28. Lagenaur C, Lemmon V. An L1-like molecule, the 8D9 antigen, is a potent substrate for neurite extension. Proc Natl Acad Sci USA 1987; 84:7753-7.
29. Springer WR, Barondes SH. Direct measurement of species-specific cohesion in cellular slime mold. J Cell Biol 1978; 78:937-42.
30. Frazier W, Glaser L. Surface components and cell recognition. Ann Rev Biochem 1979; 48:491-523.
31. McClay DR, Wessel GM, Marchase RB. Intercellular recognition: quantitation of initial binding events. Proc Natl Acad Sci USA 1981; 4975-9.
32. Lotz MM, Burdsal CA, Erickson HP et al. Cell adhesion to fibronectin and tenascin: quantitative measurements of initial bind-

ing and subsequent strengthening response. J Cell Biol 1989; 109:1795-805.
33. Benchimol S, Fuks A, Jothy S et al. Carcinoembryonic antigen, a human tumor marker, functions as an intercellular adhesion molecule. Cell 1989; 57:327-34.
34. Filbin MT, Walsh FS, Trapp JA et al. Role of myelin Po-protein as a homophilic adhesion molecule. Nature 1990; 344:871-2.
35. Furley AJ, Morton SB, Manalo D et al. The axonal glycoprotein TAG-1 is an immunoglobulin superfamily member with neurite outgrowth-promoting activity. Cell 1990; 61:157-70.
36. Edelman GM, Murray BA, Mege R, Cunningham BA, Gallin WJ. Cellular expression of liver and neural cell adhesion molecules after transfection with their cDNAs results in specific cell-cell binding. Proc Natl Acad Sci USA 1987; 84:8502-6.
37. Pizzey JA, Rowett LH, Barton CH et al. Intercellular adhesion mediated by human muscle neural cell adhesion molecule: Effects of alternative exon use. J Cell Biol 1989; 109:3465-76.
38. Kamboj RK, Lam TY, Siu CH. Regulation of slug size by the cell adhesion molecule gp80 in *Dictyostelium discoideum*. Cell Regulation 1990; 1:715-29.
39. Miura M, Asou H, Kobayashi M et al. Functional expression of a full-length cDNA coding for rat neural cell adhesion molecule L1 mediates homophilic intercellular adhesion and migration of cerebellar neurons. J Biol Chem 1992; 267:10752-8.
40. Johnson PW, Abramow-Newerly W, Seilheimer B et al. Recombinant myelin-associated glycoprotein confers neural adhesion and neurite outgrowth function. Neuron 1989; 3:377-85.
41. Gennarini G, Durbec P, Boned A et al. Transfected F3/F11 neuronal cell surface protein mediates intercellular adhesion and promotes neurite outgrowth. Neuron 1991; 6:595-606.
42. Nose A, Nagafuchi A, Takeichi M. Expressed recombinant cadherins mediate cell sorting in model system. Cell 1988; 54:993-1001.
43. Matsunaga M, Hatta K, Nagafuchi A et al. Guidance of optic nerve by N-cadherin adhesion molecules. Nature 1988; 334:62-4.
44. Doherty P, Barton CH, Dickson G, Seaton P, Rowett LH, Moore SE, Gower HJ, Walsh FS. Neuronal process outgrowth of human sensory neurons on monolayers of cells transfected with cDNAs for five human NCAM isoforms. J Cell Biol 1989; 109:789-98.
45. Kamoj RK, Siu CH. Mapping of the monoclonal antibody 80L5C4 epitope on the cell adhesion molecule gp80 of *Dictyostelium discoideum*. Biochim Biophys Acta 1988; 951:78-84.
46. Watanabe M, Frelinger III AL, Rutishauser U. Topography of NCAM structural and functional determinants. I. Classification of monoclonal antibody epitopes. J Cell Biol 1986; 103:1721-7.
47. Frelinger III AL, Rutishauser U. Topography of NCAM structural and functional determinants. II. Placement of monoclonal antibody epitopes. J Cell Biol 1986; 103:1729-37.

48. Rao Y, Wu XF, Gariepy J et al. Identification of a peptide sequence involved in homophilic binding in the neural cell adhesion molecule NCAM. J Cell Biol 1992; 118:937-49.
49. Geysen HM, Rodda SJ, Mason TJ et al. Strategies for epitope analysis using peptide synthesis. J Immunol Methods 1987; 102:259-74.
50. Kamboj RK, Gariepy J, Siu CH. Identification of an octapeptide involved in homophilic interaction of the cell adhesion molecule gp80 of *Dictyostelium discoideum*. Cell 1989; 59:615-625.
51. Staunton DE, Martin SD, Stratowa C et al. Primary structure of ICAM-1 demonstrates interaction between members of the immunoglobulin and integrin supergene family. Cell 1988; 52:925-33.
52. Obara M, Kan MS, Yamada K. Site-directed mutagenesis of the cell-binding domain of human fibronectin: separable synergistic sites mediate adhesive function. Cell 1988; 53:649-57.
53. Nose A, Tsuji K, Takeichi M. Localization of specificity determining sites in cadherin cell adhesion molecules. Cell 1990; 61:147-55.
54. Rao Y, Wu XF, Yip P et al. Structural characterization of a homophilic binding site in the neural cell adhesion molecule NCAM. J Biol Chem 1993; 268:20630-8.
55. Alzari PM, Lascombe MB, Poljak RJ. Three-dimensional structure of antibodies. Ann Rev Immunol 1988; 6:555-80.
56. Davies DR, Padlan EA, Sheriff S. Antibody-antigen complexes. Ann Rev Biochem 1990; 59:439-73.
57. Driscoll PC, Cyster JG, Campbell ID et al. Structure of domain 1 of rat T lymphocyte CD2 antigen. Nature 1991; 353:762-5.
58. Jones EY, Davis SJ, Williams AF et al. Crystal structure at 2.8 A resolution of a soluble form of the cell adhesion molecule CD2. Nature 1992; 360:232-9.

CHAPTER 3

The Structure, Function and Regulation of Cadherins

Sandra B. Munro and Orest W. Blaschuk

PREFACE

The cadherins are a rapidly expanding gene superfamily of calcium-dependent, cell adhesion molecules (CAMs). These CAMs are key regulators of morphogenesis.[1,2] In this article, we discuss the biochemical and biological properties of the cadherins. Emphasis is placed on the pivotal role of these CAMs in the establishment of epithelial cell polarity and in the maintenance of the differentiated, epithelial phenotype.

CADHERIN STRUCTURE

Members of the cadherin gene superfamily can be placed into one of nine groups (Tables 3.1 and 3.2). Most of the studies to date have focused on the classical cadherins (designated as type I cadherins), as this group of cadherins was the first to be identified.[3,4] We will initially discuss the biochemical properties of type I cadherins, and then briefly comment on the structures of the other eight groups of cadherins. The reader is invited to consult the recent reviews by Takeichi,[5] Kemler,[6] and Grunwald[7] for further information on this subject.

Cell Adhesion and Invasion in Cancer Metastasis, edited by Pnina Brodt.

Classical Cadherins

Classical (or type I) cadherins are integral membrane glycoproteins.[3,8-10] They are composed of five extracellular domains (each of which contains ~110 amino acid residues), a single transmembrane domain, and two cytoplasmic domains (Fig. 3.1). The extracellular domains contain the acidic amino acid motifs DXNDN and DXD in conserved positions. These motifs are capable of binding calcium.[11]

Table 3.1. Classification of cadherin family members

Type of cadherin	Description	Representative members	Reference
I.	Classical cadherins	E-cadherin (uvomorulin)	6,71
		P-cadherin	9
		N-cadherin	18
		B-cadherin	72
		R-cadherin (cadherin-4)	35,73
		EP-cadherin	75
II.	Atypical cadherins	OB-cadherin (cadherin-11)	76
		M-cadherin	28
		cadherin-5	35
		cadherin-8	35
		cadherin-12	35
		DE-cadherin	unpublished (acc.# D28749)
		K-cadherin	75
III.	Peptide transporters	intestinal proton-dependent peptide transporter (LI-cadherin)	77,78
IV.	Truncated cadherins	T-cadherin	27,32
V.	Protocadherins	protocadherin 43	33
VI.	Desmocollins	desmocollin 1	79
VII.	Desmogleins	desmoglein 1	80
VIII.	Receptor tyrosine kinases	c-ret	26
IX.	Others	*fat* protein	29

Table 3.2. Criteria for categorizing cadherins

Type of cadherin	Description	ICJ[a]	HAV Motif[b]	Extra EC[c]	CP length[d]	TK Domain[e]	CP Cys[f]	PKC Site[g]	PKA Site[h]	Reference
		Criteria								
I	Classical cadherin	yes	yes	no	151	no	0	1	0	81
II	Atypical cadherin	?[i]	no	no	157	no	1	1	1	82
III	Peptide transporter	no	no	yes	24	no	0	0	0	77
IV	Truncated cadherin	no	no	no	0	no	0	0	0	27
V	Protocadherin	?	no	yes	187	no	1	3	0	33
VI	Desmocollin	yes	no	no	123	no	5	5	0	83
VII	Desmoglein	yes	no	no	360	no	5	5	1	84
VIII	Receptor tyrosine kinase	no	no	yes	457	yes	5	3	1	26
IX	*Others* (fat)	?	no	yes	538	no	3	6	0	29

a. Component of intercellular junctions
b. Classical cadherins contain the cell adhesion recognition motif, His-Ala-Val[13]
c. More than 5 extracellular domains
d. Number of amino acid residues in the cytoplasmic domain
e. Tyrosine kinase domain
f. Number of Cys residues in cytoplasmic domain
g. Potential protein kinase C phosphorylation site, Ser/Thr-X-Arg/Lys[83]
h. Potential protein kinase A phosphorylation site, Arg/Lys-Arg/Lys-X-Ser/Thr[85]
i. Not determined

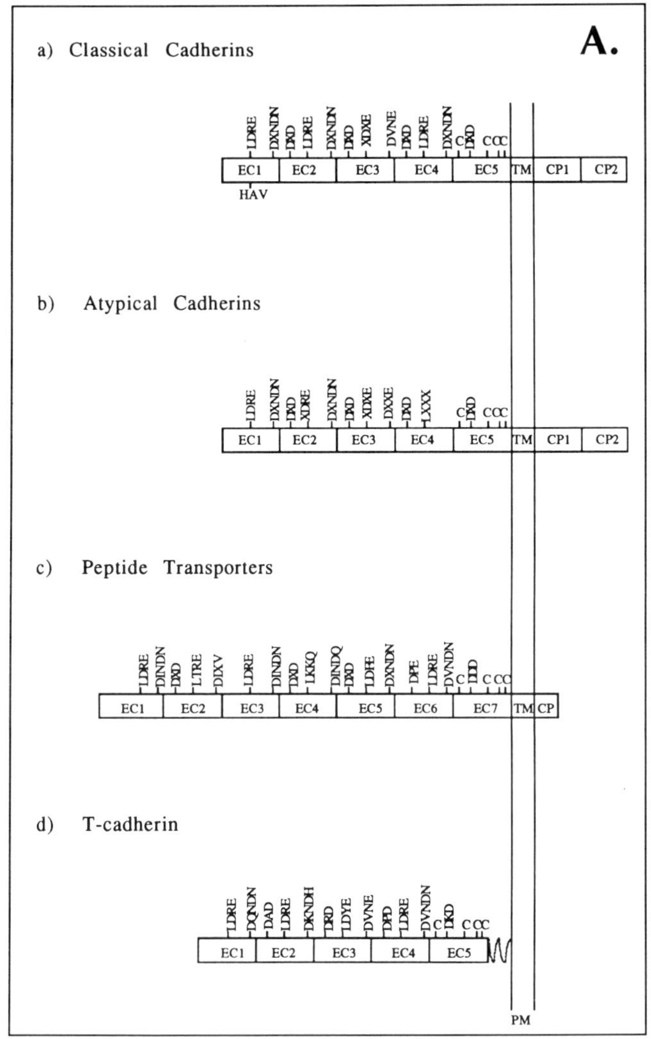

Fig. 3.1.A. Schematic diagrams illustrating the structures of members of the nine groups that comprise the cadherin gene superfamily. The structure of a classical cadherin is shown in (a). The classical cadherins have five extracellular domains (EC1-EC5), a hydrophobic domain (TM) embedded in the plasma membrane (PM), and two cytoplasmic domains (CP1 and CP2). The EC2-EC4 domains have calcium binding motifs LDRE, DXNDN and DXD (X denotes any amino acid), in conserved positions. The EC1 domain contains LDRE and DXNDN, and the EC5 domain contains DXD in conserved positions. The extracellular domain immediately adjacent to the plasma membrane (EC5) contains four cysteine residues (C) in conserved positions. The EC1 domain contains the cadherin cell adhesion recognition sequence, HAV (His-Ala-Val). The structure common to atypical cadherins is shown in (b). All the members of this group are integral membrane glycoproteins. Their structures are similar to those of the classical cadherins. They possess the cadherin calcium binding motifs and the cysteine residues adjacent to the transmembrane domain (TM). The cytoplasmic domains (CP1 and CP2) of the atypical cadherins share a high homology with those of the classical cadherins. None of the atypical cadherins harbor the CAR sequence, HAV. The structure of the intestinal peptide transporter is shown in (c). The extracellular portion of the intestinal peptide transporter is composed of seven extracellular domains (EC1-EC7), each of which contains calcium binding motifs similar to those of the classical cadherins. The domain preceding the TM contains four cysteine residues. The cytoplasmic domain (CP) of the intestinal peptide transporter is short (24 amino acids in length). It is not similar to the cytoplasmic domains of the classical cadherins. The structure of T-cadherin is shown in (d).

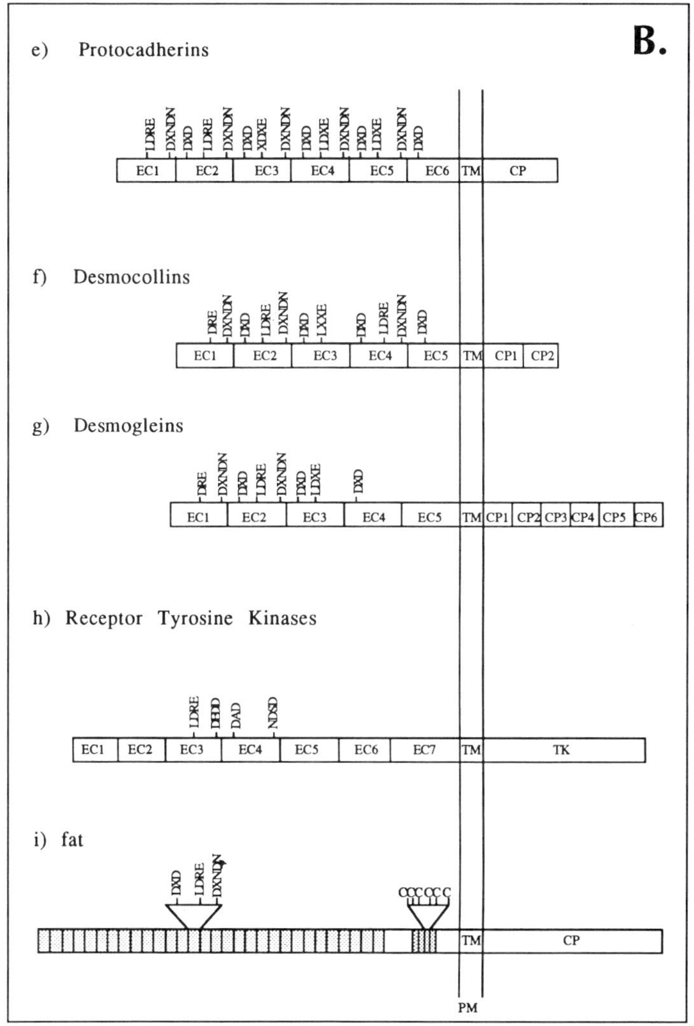

Fig. 3.1.B. T-cadherin contains five extracellular domains (EC1-EC5) which are similar to those of the classical cadherins. It differs from the classical cadherins in that it is attached to the plasma membrane via a glycosyl phosphatidylinositol anchor. The structure of a protocadherin is shown in (e). Protocadherins have at least 6 cadherin-like extracellular domains (EC1-EC6). In contrast, their cytoplasmic domains (CP) differ from those of the other types of cadherins. The structure common to desmocollins I and II is shown in (f). The desmocollins are integral membrane glycoproteins with DXNDN, DXD, and LDRE motifs present in positions corresponding to those found in the classical cadherins. The cytoplasmic domains (CP1 and CP2) are not homologous to those of the cadherins. The structure of desmoglein is shown in (g). The extracellular portion of desmoglein shares homology with the cadherins. In contrast, it has many more cytoplasmic domains (CP1-CP6) than the other cadherins. The structure of c-ret is shown in (h). It has an intracellular tyrosine kinase domain (TK) and seven extracellular domains (EC1-EC7). c-ret contains repeated calcium binding motifs (DEDD, DAD, NDSD) that are similar in sequence and position to those of the classical cadherins. One LDRE motif is present in the third extracellular domain (EC3). In addition, c-ret has a cysteine-rich domain (containing 15 cysteine residues; EC7) adjacent to its transmembrane domain (TM). The structure of the protein encoded by the Drosophila *tumor suppressor gene fat is shown in (i). The fat protein contains 34 cadherin-like domains, each containing the DXD, LDRE, and DXNDN motifs. The extracellular portion also contains four EGF-like domains, each of which harbors six cysteine residues in conserved positions. There is a hydrophobic transmembrane domain (TM), followed by a large cytoplasmic domain (CP). The cytoplasmic domain has no homology to the cytoplasmic domains of the classical cadherins.*

The first extracellular domain of each classical cadherin harbors the cell adhesion recognition (CAR) sequence, HAV.[12,13] Either synthetic peptides containing the CAR sequence, or antibodies directed against the CAR sequence are capable of inhibiting cadherin-dependent processes.[12,14-17] For example, the synthetic peptides LRAHAVDVNG-amide and LRAHAVDING-amide (the amino acid sequences of these two peptides are identical to those of the HAV-containing regions of avian and murine N-cadherin, respectively) are capable of blocking neurite outgrowth, a process mediated by N-cadherin.[13,15] In comparison, the synthetic peptides, LRAHAIDING-amide and LRARAVDING-amide are less potent inhibitors of neurite outgrowth on astrocytes, whereas the decamer, LRAHGVDING-amide is incapable of inhibiting this process (personal observations). These observations suggest that the tripeptide, HAV is directly involved in regulating cadherin-mediated cell adhesion.

Although cadherins are believed to promote cell adhesion through calcium-dependent, homophilic interactions,[3,10,18] several studies have shown that these CAMs are also capable of forming heterotypic complexes with one another under certain circumstances (Fig. 3.2).[4,10,19] The adhesive affinity of heterotypic cadherin complexes is lower than that of homotypic cadherin complexes. Cadherins that have been altered by introducing amino acid substitutions into the region surrounding the CAR sequence do not preferentially form homotypic complexes over heterotypic complexes.[20] The nonconserved amino acid residues immediately adjacent to the HAV site are thus thought to be involved in modulating the ability of cadherins to interact with one another in a homotypic manner.[20]

The cytoplasmic domains are the most highly conserved regions of the classical cadherins.[18] These domains are associated with three intracellular proteins, known as α-, β- and γ-catenin.[5,6,21] The catenins are thought to mediate the interaction between the cadherins and the microfilaments of the cytoskeleton. Remarkably, the cadherins cannot promote cell adhesion unless they are complexed with the catenins.[22-25] Furthermore, posttranslational modifications of the catenins affect the ability of the cadherins to mediate cell adhesion.[6] For example, the phosphorylation of β-catenin by pp60$^{v\text{-}src}$ leads to a decrease in the strength of cell adhesion. The catenins can therefore be regarded as important regulators of cadherin function.

CADHERIN TYPES II-IX

The eight other groups of cadherins (designated as cadherin types II-IX, respectively) represent a varied collection of membrane glycoproteins (Table 3.1 and Fig. 3.1). All members of these eight groups possess the calcium-binding motifs that are present in the classical cadherins, but none of them possess the CAR sequence, HAV.[26-31] The CAR sequences of the cadherin types II-IX have not been ascertained. Furthermore, most of the cadherin types II-IX are incapable of interacting with catenins.[27,30-33] Finally, the ability of most of these cadherins to promote cell adhesion has not been determined. Studies have shown that the protocadherins,[33] desmogleins,[34] and several atypical cadherins[35] do not mediate cell adhesion in a manner similar to that of the classical cadherins. The biological functions of the cadherin types II-IX therefore remain uncertain.

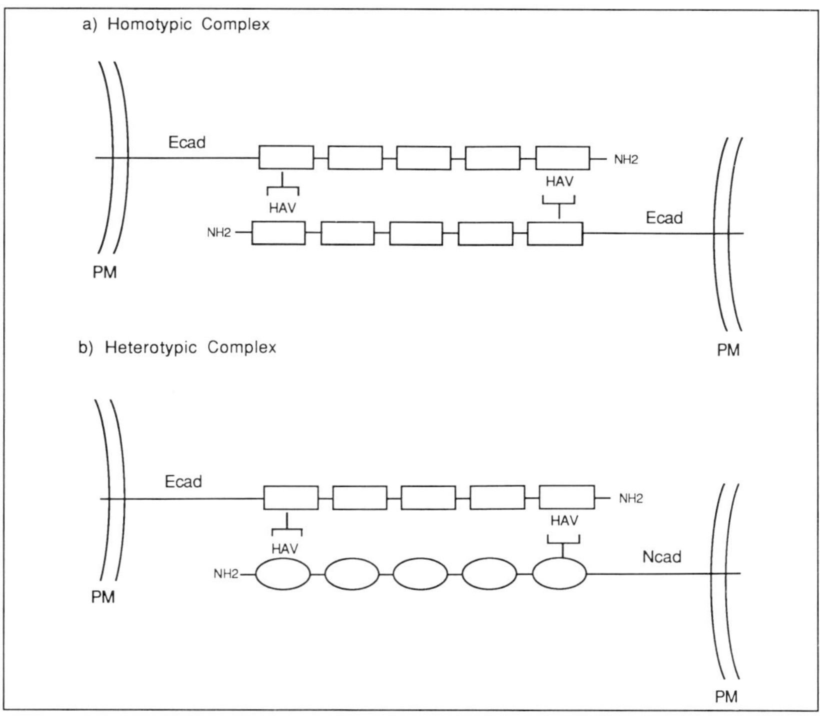

Fig. 3.2. Schematic diagram illustrating homotypic (a) and heterotypic (b) cadherin complexes. In (a), E-cadherin (E-cad) is shown transversing the plasma membrane (PM) of a cell and forming a homotypic complex with another E-cad expressed by an adjoining cell. In (b), E-cad is shown interacting with N-cadherin (N-cad) to form a heterotypic complex. Both types of interactions are mediated by the cell adhesion recognition sequence, HAV (His-Ala-Val). Although cadherins are capable of forming both types of complexes, homophilic interactions are favoured over heterophilic interactions.

DISTRIBUTION OF CLASSICAL CADHERINS IN ADULT TISSUES

Among the best characterized of the classical cadherins are the three murine CAMs: E-cadherin (E-cad; M_r~124 kDa), N-cadherin (N-cad; M_r~130 kDa), and P-cadherin (P-cad; M_r~118 kDa) (reviewed by Takeichi;[3] Geiger and Ayalon[4]). In the adult animal, these cadherins have been localized to the membrane domains of adherens junctions (Fig. 3.3).[4] The distribution of classical cadherins differs within adult vertebrate tissues.[36] For example, murine E-cad is expressed by epithelial cells, but not by most neural or muscle cells. N-cad is expressed by neural and muscle cells, but not by most epithelial cells. The only tissues in which murine P-cad is present in abundance are the placenta and uterine decidua.

DEVELOPMENTAL BIOLOGY OF THE CLASSICAL CADHERINS

The spatiotemporal expression of the cadherins is strictly regulated during development.[36] E-cad is first expressed in mouse embryos at the one cell stage.[37-39] P-cad is first detectable in the mural

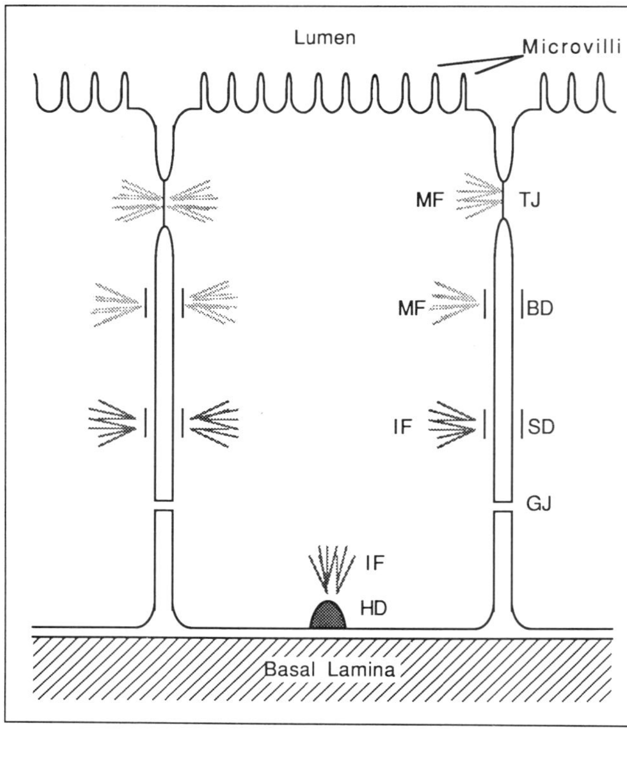

Fig. 3.3. Schematic diagram illustrating the intercellular junctions formed by epithelial cells. Tight junctions (TJ or zonula occludens*) form a barrier which limits the movement of substances from one tissue compartment to another. Belt desmosomes (BD or* zonula adherens*) form a continuous band around adjacent cells and lend structural integrity to epithelial tissues. Classical cadherins are localized to the membrane domains of these junctions. Microfilaments (MF) of the cytoskeleton are associated with both TJ and BD. Spot desmosomes (SD or* macula adherens*) are punctate junctions at which intermediate filaments (IF) are anchored to the plasma membrane. These intercellular junctions also provide structural integrity to the epithelium. Desmogleins and desmocollins are localized to the membrane domain of SD. Gap junctions (GJ) are transmembrane channels which allow the exchange of inorganic ions and small metabolites between the cytoplasms of adjoining cells. Hemidesmosomes (HD) mediate the attachment of epithelial cells to the basal lamina.*

trophectoderm of 4.5 day blastocysts,[40] whereas N-cad is initially expressed by the mesodermal cells of the gastrula.[36]

The expression of the classical cadherins by the different cell types of the developing embryo appears to be complementary.[36] Cells displaying different cadherin subtypes are capable of segregating from one another in vitro.[20] The expression of different cadherin subtypes by the cells of the three germ layers therefore provides a molecular basis for the segregation of embryonic populations.

E-cad has been shown to mediate the calcium-dependent compaction of mouse blastomeres at the eight-cell stage of embryogenesis.[37-39] Furthermore, mouse embryos homozygously negative for E-cad fail to form a trophectodermal epithelium.[37] E-cad thus plays a pivotal role in the biogenesis of the mammalian epithelium (see Sections 4.1 and 4.2 for further discussion regarding this issue).

The role of N-cad in vertebrate development has also been extensively investigated.[41-43] For example, the microinjection of mRNA encoding this CAM into *Xenopus* oocytes causes N-cad to be prematurely, ectopically expressed by the early ectodermal cells of the developing embryos.[41] The misexpression of N-cad results in severe developmental abnormalities, such as failure of the neural tube to close, and in some cases, the development of embryos lacking heads and tails.[41] N-cad therefore appears to be an important regulator of early morphogenetic events.

THE ROLE OF CLASSICAL CADHERINS IN THE ESTABLISHMENT OF CELL JUNCTIONS

In addition to regulating the formation of adherens junctions, the classical cadherins are also capable of mediating the establishment of other intercellular junctions.[44-46] For example, antibodies directed against E-cad inhibit the normal formation of tight junctions, belt desmosomes, and spot desmosomes by Madin-Darby canine kidney (MDCK) cells (Fig. 3.3).[44] In another study, Mege et al[45] showed that mouse sarcoma cells transfected with cDNA encoding L-CAM (the avian equivalent of murine E-cad) form gap junctions and belt desmosomes when grown to confluence. In contrast, untransfected sarcoma cells, as well as cells transfected with cDNA encoding N-CAM (a calcium-independent CAM), were found to be incapable of forming intercellular junctions with one another at confluence. These results demonstrate the capability of cadherins to promote the establishment of intercellular junctions.

The Role of Classical Cadherins in the Establishment of Cell Polarity

In view of the classical cadherins' ability to mediate the formation of intercellular junctions, it follows that these CAMs are also capable of inducing epithelial cell polarity.[47-53] For example, E-cad has been shown to be capable of regulating the distribution of sodium, potassium-adenosine triphosphatase (Na^+,K^+-ATPase)[47,48] and the epidermal growth factor receptor (EGF-R) within the plane of the epithelial cell plasma membrane.[51,52] McNeill et al[47] demonstrated that the distribution of Na^+,K^+-ATPase becomes restricted to sites of E-cad-mediated cell contact when cadherin-deficient fibroblasts are transfected with cDNA encoding E-cad. Similarly, Suarez-Quian and Byers[51] showed that E-cad-mediated cell adhesion causes the EGF-R to become localized to sites of cell contact.

E-cad cannot modulate the plasma membrane distribution of either Na^+, K^+-ATPase or the EGF-R unless it is complexed with the catenins.[47,52] In view of these observations, Nathke et al[50] and Hoschuetzky et al[52] speculate that the catenins function as intracellular regulators of E-cad action. Hoschuetzky et al[52] have demonstrated that β-catenin can bind to the EGF-R. This catenin may therefore be a facilitator of complex formation between E-cad and other plasma membrane proteins, such as the EGF-R.

The Role of E-Cadherin in the Maintenance of the Epithelial Cell State

In addition to regulating the formation of intercellular junctions and the establishment of epithelial cell polarity, E-cad has also been implicated as a morphogenetic regulator of the well-differentiated, non-invasive epithelial cell state.[54-57] For example, antibodies directed against this CAM prevent the formation of intercellular junctions between MDCK cells (Fig. 3.3), thereby altering the morphology of these cells (from epithelial to mesenchymal) and stimulating their ability to invade collagen gels.[54] The loss of E-cad-mediated cell adhesion is currently thought to initiate a series of events culminating in the transformation of epithelial cells into migratory and invasive fibroblast-like cells. Several studies concerning the analysis of E-cad levels in human carcinoma cell lines support this hypothesis (reviewed by Birchmeier et al;[55] Behrens[56]). In particular, Frixen et al[57] have shown that non-invasive carcinoma cell lines with an epithelioid phenotype express E-cad, whereas poorly-differentiated, invasive carcinoma cell lines tend not to express this CAM. Finally, it is noteworthy that invasive, fibroblast-like carcinoma cells become non-invasive when transfected with cDNA encoding E-cad. Together, these findings illustrate the ability of E-cad to act as a morphogenetic regulator.

EXPRESSION OF E-CAD IN HUMAN CARCINOMAS

In the preceding section, we presented data suggesting that the down-regulation of E-cad promotes the invasiveness of carcinoma cells in vitro. Studies have now emerged which reveal that human carcinomas also contain reduced E-cad levels in vivo (reviewed by Takeichi;[5] Behrens;[56] Giroldi and Schalken[58]). In particular, poorly differentiated carcinomas have been shown to contain either reduced, or undetectable levels of E-cad, whereas well-differentiated tumors invariably display this CAM. The down-regulation of E-cad therefore appears to facilitate the tumorigenic process. Birchmeier et al[55] and others[54,59,60] contend that the gene encoding E-cad is a tumor suppressor gene.

A number of different mechanisms may lead to the down-regulation of E-cad in cancer cells. For example, studies have shown that alleles are frequently lost from chromosome 16q in human prostate and breast cancers, as well as in hepatocellular carcinoma and Wilms' tumors.[61-64] The E-cad gene resides in this chromosomal region (16q22.1).[65] The lower E-cad levels frequently observed in poorly differentiated cancer cells may therefore be a consequence of reduced E-cad gene dosage due to allelic loss. Unfortunately, this relatively straightforward hypothesis is insufficient to explain all aspects of E-cad expression by cancer cells. For example, only 30% of prostate tumors exhibit a loss of heterozygosity on chromosome 16q.[63] Furthermore, E-cad-positive rat prostate cancer cells have been reported to occur in metastatic lung deposits derived from anaplastic E-cad-negative primary tumor cells.[66] E-cad levels may thus be only transiently suppressed in some tumor cells. The transient suppression of classical cadherins has also been observed to occur during various phases of vertebrate development (for example, during the formation of the peripheral ganglia).[67] Finally, it should be noted that carcinoma cells displaying reduced E-cad levels are invasive.[59] The total elimination of E-cad expression is therefore not a necessary prerequisite for carcinoma cells to become invasive.[59] As tumors are usually composed of a heterogeneous population of cells displaying a wide spectrum of E-cad levels,[5] it becomes imperative to identify factors that are capable of locally regulating cadherin levels.

STEROIDS AS REGULATORS OF CLASSICAL CADHERIN LEVELS

We have shown that steroids are capable of stimulating cadherin levels in vitro and in vivo.[68-70] In particular, we have found that estradiol has the ability to influence E-cad and N-cad levels in all of the systems that we have examined to date (rat ovarian granulosa cells, human breast carcinoma cells, mouse testis, mouse ovary and mouse uterus). Based on these observations, we have proposed the hypothesis that steroids are the major regulators of vertebrate cadherin levels. We speculate that the multiple morpho-regulatory actions exerted by steroids on developing tissues may arise from their ability to modulate cadherin levels. Steroid-induced alterations in cadherin levels would have dramatic effects on morphogenetic processes (including tumorigenesis), as these CAMs mediate the establishment of functional cell collectives.[1,2]

SUMMARY AND FUTURE DIRECTIONS

We have discussed the ability of the classical cadherins to act as morphogenetic regulators. Studies have revealed that these CAMs exert their effects by promoting cell adhesion, the formation of intercellular junctions, and the establishment of cell polarity. The cadherins (in particular, E-cad) have also been shown to play a key role in maintaining the differentiated, non-invasive epithelial cell state.

Many major issues concerning cadherin function remain unexplored. The need to identify factors that are capable of regulating cadherin levels in tissues has become apparent. Although steroids have been implicated as regulators of E-cad and N-cad in vitro and in vivo, the mechanisms by which these hormones act are unknown. Further, the biological functions of the other cadherin types (such as the atypical cadherins and protocadherins) must be determined. Finally, the intracellular mechanisms involved in regulating cadherin function remain to be fully elucidated. The resolution of these multiple issues is essential, if we are to fully appreciate the biological properties of the cadherins.

ACKNOWLEDGMENTS

The authors express their gratitude to the Medical Research Council of Canada and the Kidney Foundation of Canada for financial support. We also thank K.L. Blaschuk for valuable comments regarding this manuscript. S.B.M. is the recipient of fellowships from the Royal Victoria Hospital Research Institute and the Division of Urology, Department of Surgery, McGill University.

REFERENCES

1. Takeichi M. Cadherin cell adhesion receptors as a morphogenetic regulator. Science 1991; 251:1451-5.
2. Edelman GM. Morphoregulation. Develop Dynamics 1992; 193:2-10.
3. Takeichi M. Cadherins: a molecular family important in selective cell-cell adhesion. Ann Rev Biochem 1990; 59:237-52.
4. Geiger B, Ayalon O. Cadherins. Annu Rev Cell Biol 1992; 8:307-32.
5. Takeichi M. Cadherins in cancer: implications for invasion and metastasis. Curr Opin Cell Biol 1993; 5:806-11.
6. Kemler R. From cadherins to catenins: cytoplasmic protein interactions and regulation of cell adhesion. Trends Genet 1993; 9:317-21.
7. Grunwald GB. The structural and functional analysis of cadherin calcium-dependent cell adhesion molecules. Curr Opin Cell Biol 1993; 5:797-805.
8. Nagafuchi A, Shirayoshi Y, Okazaki K et al. Transformation of cell adhesion properties by exogenously introduced E-cadherin cDNA. Nature 1987; 329:341-3.
9. Nose A, Nagafuchi A, Takeichi M. Isolation of placental cadherin cDNA: identification of a novel gene family of cell-cell adhesion molecules. EMBO J 1987; 6:3655-61.
10. Miyatani S, Shimamura K, Hatta M et al. Neural cadherin: role in selective cell-cell adhesion. Science 1989; 245:631-5.
11. Ozawa M, Engel J, Kemler R. Single amino acid substitutions in one calcium-binding site of uvomorulin abolish the adhesive function. Cell 1990; 63:1033-8.
12. Blaschuk OW, Pouliot Y, Holland PC. Identification of a conserved region common to cadherins and influenza strain A hemagglutinins. J Mol Biol 1990; 211:679-82.
13. Blaschuk OW, Sullivan R, David S et al. Identification of a cadherin cell adhesion recognition sequence. Develop Biol 1990; 139:227-9.

14. Chuah MI, David S, Blaschuk OW. Differentiation and survival of rat olfactory epithelial neurons in dissociated cell culture. Develop Brain Res 1991; 60:123-32.
15. Doherty P, Rowett LH, Moore SE et al. Neurite outgrowth in response to transfected N-CAM and N-cadherin reveals fundamental differences in neuronal responsiveness to CAMs. Neuron 1991; 6:247-58.
16. Alexander JS, Blaschuk OW, Haselton FR. An N-cadherin-like protein contributes to solute barrier maintenance in cultured endothelium. J Cell Physiol 1993; 156:610-18.
17. Newton SC, Blaschuk OW, Millette CF. N-cadherin mediates Sertoli cell-spermatogenic cell adhesion. Develop Dynam 1993; 197:1-13.
18. Hatta K, Nose A, Nagafuchi A et al. Cloning and expression of cDNA encoding a neural calcium-dependent cell adhesion molecule: its identity in the cadherin gene family. J Cell Biol 1988; 106:873-81.
19. Volk T, Cohen O, Geiger B. Formation of heterotypic adherens-type junctions between L-CAM-containing liver cells and A-CAM-containing lens cells. Cell 1987; 50:987-94.
20. Nose A, Tsuji K, Takeichi M. Localization of specificity determining sites in cadherin cell adhesion molecule. Cell 1990; 61:147-55.
21. Hinck L, Nathke IS, Papkoff J et al. Dynamics of cadherin/catenin complex formation: novel protein interactions and pathways of complex assembly. J Cell Biol 1994; 125:1327-40.
22. Nagafuchi A, Takeichi M; Cell binding function of E-cadherin is regulated by the cytoplasmic domain. EMBO J 1988; 7:3679-84.
23. Ozawa M, Ringwald M, Kemler R. Uvomorulin-catenin complex formation is regulated by a specific domain in the cytoplasmic region of the cell adhesion molecule Proc Natl Acad Sci USA 1990; 87:4246-50.
24. Shimoyama Y, Nagafuchi A, Fujita S et al. Cadherin dysfunction in a human cancer cell line: possible involvement of loss of α-catenin expression in reduced cell-cell adhesiveness. Cancer Res 1992; 52:5770-4.
25. Hirano S, Kimoto N, Shimoyama Y et al. Identification of a neural α-catenin as a key regulator of cadherin function and multicellular organization. Cell 1992; 70:293-301.
26. Schneider R. The human protooncogene *ret*: a communicative cadherin? Trends Biochem Sci 1992; 17:468-9.
27. Ranscht B, Dours-Zimmermann MT. T-cadherin, a novel cadherin cell adhesion molecule in the nervous system lacks the conserved cytoplasmic region. Neuron 1991; 7:391-402.
28. Donalies M, Cramer M, Ringwald M et al. Expression of M-cadherin, a member of the cadherin multigene family, correlates with differentiation of skeletal muscle cells. Proc Natl Acad Sci USA 1991; 88:8024-8.

29. Mahoney PA, Weber U, Onofrechuk P et al. The *fat* tumor suppressor gene in *Drosophila* encodes a novel member of the cadherin gene superfamily. Cell 1991; 67:853-68.
30. Collins JE, Legan PK, Kenny TP et al. Cloning and sequence analysis of desmosomal glycoproteins 2 and 3 desmocollins: cadherin-like desmosomal adhesion molecules with heterogenous cytoplasmic domains. J Cell Biol 1991; 113:381-91.
31. Goodwin L, Hill JE, Raynor K et al. Desmoglein shows extensive homology to the cadherin family of cell adhesion molecules. Biochem Biophys Res Commun 1990:1224-30.
32. Vestal DJ, Ranscht B. Glycosyl phosphatidylinositol anchored T-cadherin mediates calcium-dependent, homophilic cell adhesion. J Cell Biol 1992; 119:451-61.
33. Sano K, Tanihara H, Heimark RL et al. Protocadherins: a large family of cadherin-related molecules in the central nervous system. EMBO J 1993; 12:2249-56.
34. Amagai M, Karpati S, Klaus-Kovtun V et al. Extracellular domain of pemphigus vulgaris antigen desmoglein 3 mediates weak homophilic adhesion. J Invest Dermatol 1994; 102:402-8.
35. Tanihara H, Kido M, Obata S et al. Characterization of cadherin-4 and cadherin-5 reveals new aspects of cadherins. J Cell Sci 1994; 107:1697-1704.
36. Takeichi M. The cadherins: cell-cell adhesion molecules controlling animal morphogenesis. Development 1988; 102:639-55.
37. Larue L, Ohsugi M, Hirchenhain J et al. E-cadherin null mutant embryos fail to form a trophectoderm epithelium. Proc Natl Acad Sci USA 1994; 91:8263-7.
38. Ogou S, Okada TS, Takeichi M. Cleavage stage mouse embryos share a common cell adhesion system with teratocarcinoma cells. Develop Biol 1982; 92:521-8.
39. Yoshida-Noro C, Suzuki N, Takeichi M. Molecular nature of the calcium-dependent cell-cell adhesion system in mouse teratocarcinoma and embryonic cells studied with a monoclonal antibody. Devel Biol 1984; 101:19-27.
40. Kadokawa Y, Fuketa I, Nose A et al. Expression pattern of E- and P-cadherin in mouse embryos and uteri during the preimplantation period. Develop Growth and Differ 1989; 31:23-30.
41. Heasman J, Crawford A, Goldstone A et al. Overexpression of cadherins and underexpression of β-catenin inhibit dorsal mesoderm induction in early *Xenopus* embryos. Cell 1994; 79:791-803.
42. Bitzur S, Kam Z, Geiger B. Structure and distribution of N-cadherin in developing zebrafish embryos: morphogenetic effects of ectopic over-expression. Develop Dynamics 1994; 201:121-36.
43. Detrick RJ, Dickey D, Kintner CR. The effects of N-cadherin misexpression on morphogenesis in *Xenopus* embryos. Neuron 1990; 4:493-506.

44. Gumbiner B, Stevenson B, Grimaldi A. The role of the cell adhesion molecule uvomorulin in the formation and maintenance of the epithelial junctional complex. J Cell Biol 1988; 107:1575-87.
45. Mege RM, Matsuzaki F, Gallin et al. Construction of epithelioid sheets by transfection of mouse sarcoma cells with cDNAs for chicken cell adhesion molecules. Proc Natl Acad Sci USA 1988; 85:7274-78.
46. Matsuzaki F, Mege R-M, Jaffe SH et al. cDNAs of cell adhesion molecules of different specificity induce changes in cell shape and border formation in cultured S180 cells. J Cell Biol 1990; 110:1239-52.
47. McNeill H, Ozawa M, Kemler R et al. Novel function of the cell adhesion molecule uvomorulin as an inducer of cell surface polarity. Cell 1990; 62:309-16.
48. Watson AJ, Damsky CH, Kidder GM. Differentiation of an epithelium: factors affecting the polarized distribution of sodium potassium-ATPase in mouse trophectoderm. Dev Biol 1990; 141:104-14.
49. Marrs JA, Napolitano EW, Murphy-Erdosh C et al. Distinguishing roles of the membrane-cytoskeleton and cadherin mediated cell-cell adhesion in generating different Na^+,K^+-ATPase distributions in polarized epithelia. J Cell Biol 1993; 123:149-64.
50. Nathke IS, Hinck L, Swedlow JR et al. Defining interactions and distributions of cadherin and catenin complexes in polarized cells. J Cell Biol 1994; 125:1341-52.
51. Suarez-Quian CA, Byers SW. Redistribution of epidermal growth factor receptor as a function of cell density, cell adhesion and calcium in human A-431 cells. Tissue and Cell 1993; 25:1-17.
52. Hoschuetzky H, Aberle H, Kemler R. β-Catenin mediates the interaction of the cadherin-catenin complex with epidermal growth factor receptor. J Cell Biol 1994; 127:1375-80.
53. Rodriguez-Boulan E, Powell SK. Polarity of epithelial and neuronal cells. Annu Rev Cell Biol 1992; 8:395-427.
54. Behrens J, Mareel MM, van Roy FM et al. Dissecting tumor cell invasion: epithelial cells aquire invasive properties after the loss of uvomorulin-mediated cell-cell adhesion. J Cell Biol 1989; 108:2435-47.
55. Birchmeier W, Behrens J, Weidner KM et al. Dominant and recessive genes involved in tumor cell invasion. Curr Opin Cell Biol 1991; 3:832-40.
56. Behrens J. The role of cell adhesion molecules in cancer invasion and metastasis. Breast Cancer Res Treat 1993; 24:175-84.
57. Frixen UH, Behrens J, Sachs M et al. E-cadherin-mediated cell-cell adhesion prevents invasiveness of human carcinoma cells J Cell Biol 1991; 111:173-85.
58. Giroldi LA, Schalken JA. Decreased expression of the intercellular adhesion molecule E-cadherin in prostate cancer: biological signifi-

cance and clinical implications Cancer Metastasis Rev 1993; 12:29-37.
59. Vleminckx K, Vakaet Jr L, Mareel M. Genetic manipulation of E-cadherin expression by epithelial tumor cells reveals an invasion suppressor role. Cell 1991; 66:107-19.
60. Hedrick L, Cho KR, Vogelstein B. Cell adhesion molecules as tumor suppressors. Trends Cell Biol 1993; 3:36-9.
61. Sato T, Tanigami A, Yamakawa K et al. Allelotype of breast cancer: cumulative allele losses promote tumor progression in primary breast cancer. Cancer Res 1990; 50:7184-89.
62. Tsuda H, Zhang W, Shimosato Y et al. Allele loss on chromosome 16 associated with progression of human hepatocellular carcinoma. Proc Natl Acad Sci USA 1990; 87:6791-4.
63. Carter BS, Ewing CM, Ward WS et al. Allelic loss of chromosomes 16q and 10q in human prostate cancer. Proc Natl Acad Sci USA 1990; 87:8751-5.
64. Maw MA, Grundy PE, Millow LJ et al. A third Wilms' tumor locus on chromosome 16. Cancer Res 1992; 52:3094-8.
65. Mansouri A, Spurr N, Goodfellow PN et al. Characterization and chromosomal localization of the gene encoding the human cell adhesion molecule uvomorulin. Differentiation 1988; 38:67-71.
66. Bussemakers MJG, Moorselaar RJA van, Giroldi LA et al. Decreased expression of E-cadherin in the progression of rat prostatic cancer. Cancer Res 1992; 52:2916-22.
67. Ranscht B. Cadherin cell adhesion molecules in vertebrate neural development. Sem Neurosci 1991; 3:285-96.
68. MacCalman CD, Blaschuk OW. Gonadal steroids regulate N-cadherin mRNA levels in the mouse testis. Endocrine 1994; 2:157-63.
69. MacCalman CD, Farookhi R, Blaschuk OW. Estradiol and progesterone regulate E-cadherin mRNA levels in the mouse uterus. Endocrine 1994; 2:485-90.
70. MacCalman CD, Farookhi R, Blaschuk OW. Estradiol regulates E-cadherin mRNA levels in the surface epithelium of the mouse ovary. Clin Exp Metastasis 1994; 12:276-82.
71. Ringwald M, Schuh R, Vestweber D et al. The structure of cell adhesion molecule uvomorulin. Insights into the molecular mechanism of Ca^{2+}-dependent cell adhesion. EMBO J 1987; 6:3647-53.
72. Napolitano EW, Venstrom K, Wheeler EF et al. Molecular cloning and characterization of B-cadherin, a novel chick cadherin. J Cell Biol 1991; 113:893-905.
73. Inuzaka H, Miyatani S, Takeichi M. R-cadherin: a novel calcium-dependent cell-cell adhesion molecule expressed in the retina. Neuron 1991; 7:69-79.
74. Ginsburg D, DeSimone D, Geiger B. Expression of a novel cadherin EP-cadherin in unfertilized eggs and early Xenopus embryos. Development 1991; 111:315-25.

75. Xiang Y-Y, Tanaka M, Suzuki M et al. Isolation of a complementary DNA encoding K-cadherin, a novel rat cadherin preferentially expressed in fetal kidney and kidney carcinoma. Cancer Res 1994; 54:3034-41.
76. Okazaki M, Takeshita S, Kawai S et al. Molecular cloning and characterization of OB-cadherin, a new member of the cadherin family expressed in osteoblasts. J Biol Chem 1994; 269:12092-8.
77. Dantzig AH, Hoskins J, Tabas LB et al. Association of intestinal peptide transport with a protein related to the cadherin superfamily. Science 1994; 264:430-3.
78. Berndorff D, Gessner R, Kreft B et al. Liver-intestine cadherin: Molecular cloning and characterization of a novel $Ca2^{+}$-dependent cell adhesion molecule expressed in liver and intestine. J Cell Biol 1994; 125:1353-69.
79. Legan PK, Yue KKM, Chidgey MAJ et al. The bovine desmocollin family: A new gene and expression patterns reflecting epithelial cell proliferation and differentiation. J Cell Biol 1994; 126:507-18.
80. Buxton RS, Cowin P, Franke WW et al. Nomenclature of the desmosomal cadherins. J Cell Biol 1993; 121:481-3.
81. Rimm DL, Morrow JS. Molecular cloning of human E-cadherin suggests a novel subdivision of the cadherin family. Biochem Biophys Res Comm 1994; 200:1754-61.
82. Tanihara H, Sano K, Heimark RL et al. Cloning of five human cadherins clarifies characteristic features of cadherin extracellular domain and provides further evidence for two structurally different types of cadherin. Cell Adhesion Commun 1994; 2:15-26.
83. Mechanic S, Raynor K, Hill JE et al. Desmocollins form a subset of the cadherin family of cell adhesion molecules. Proc Natl Acad Sci USA 1991; 88:4476-80.
84. Amagai M, Klaus-Kovtun V, Stanley JR. Autoantibodies against a novel epithelial cadherin in pemphigus vulgaris, a disease of cell adhesion. Cell 1991; 67:869-77.
85. Bolander FF. Molecular Endocrinology. San Diego: Academic Press 1989:93.

CHAPTER 4

The Integrins: Mediators of Cell-extracellular Matrix and Intercellular Communication

Pnina Brodt and Shoukat Dedhar

INTRODUCTION

Cell communication with the extracellular matrix, (ECM) is a central regulatory mechanism in a wide range of cellular processes including cell adhesion, spreading, migration, proliferation and differentiation. The major mediators of this communication are members of the integrin family of adhesion receptors. Each member of this family is an α/β heterodimer with a distinct ligand-specificity conferred by the individual subunits.[1]

The study of integrins, which began in the seventies, has undergone a rapid expansion in the past decade. During that period it has become clear that earlier definitions of these receptors as linkers between the extracellular matrix proteins and the cellular cytoskeleton, whose main function is to induce changes in cell shape and motility, were simplistic and limited. More recent evidence revealed that the integrins can play a major role in transducing a range of signals from the extracellular microenvironment into the cell and can act in a manner akin to growth factor receptors. In so doing they can trigger a chain of molecular events leading to profound changes in cell behavior.[2]

Cell Adhesion and Invasion in Cancer Metastasis, edited by Pnina Brodt.

Several excellent reviews dealing with all aspects of integrin structure and function have been published in recent years.[2-7] It is not the objective of this chapter to provide another comprehensive review of the integrin literature but rather to summarize sufficient background information for the main focus of this book namely, the role of cellular adhesion in cancer metastasis. It is also with this objective in mind that the role of integrins in signaling, gene activation and growth regulation will be highlighted.

THE INTEGRIN SUBFAMILIES

To date, 15 α subunits and 8 β subunits have been identified. Each subunit is a transmembrane glycoprotein consisting of an extracellular domain, a membrane-spanning region and a cytoplasmic domain which, with the exception of the β_4 subunit is relatively short (40-60 amino acids). The different α and β subunits combine to form at least 21 different non-covalently associated heterodimers which have been subdivided on the basis of a shared β subunit into three major subfamilies: the β_1, β_2 and β_3 families.[3,8] The other β chains, β_4-β_8 and β_P, appear to associate each with one α subunit only (see Table 4.1).

THE β_1 FAMILY

Members of the β_1 subfamily, also known as VLA integrins (Very Late Antigens of T cell activation) are expressed on a wide range of cell types and are probably present in all animal species.[3] To date, 10 heterodimers have been identified in this family, all involved in recognition of one or more extracellular matrix proteins. These include fibronectin (FN) which is recognized by integrins $\alpha_5\beta_1$ (the major FN receptor), $\alpha_3\beta_1$, $\alpha_4\beta_1$, $\alpha_2\beta_1$ and $\alpha_v\beta_1$, laminin which is recognized by integrins $\alpha_1\beta_1$, $\alpha_2\beta_1$, $\alpha_6\beta_1$, $\alpha_7\beta_1$ and possibly $\alpha_3\beta_1$, collagens which are recognized by $\alpha_1\beta_1$, $\alpha_2\beta_1$, and possibly $\alpha_3\beta_1$, vitronectin which is recognized by $\alpha_v\beta_1$ and epiligrin, an epithelial basement membrane protein which is recognized by $\alpha_3\beta_1$ (reviewed in refs. 3, 8 and 9). Thus several different receptors may recognize more than one matrix protein and each of these proteins can be recognized by multiple receptors, providing a large pool of receptors for cell-ECM communication. In addition, integrin $\alpha_4\beta_1$ and possibly α_2 and $\alpha_3\beta_1$ can also mediate cell-cell adhesion, $\alpha_4\beta_1$ through recognition of VCAM-1—a member of the Ig supergene family[10,11] and $\alpha_3\beta_1$ through a determinant yet to be identified possibly a cell associated $\alpha_2\beta_1$.[3,12,13] Although all members of this group have a common β subunit, their ligand specificities are diverse and distinct. Only 3 members of this subfamily $\alpha_3\beta_1$, $\alpha_5\beta_1$ and $\alpha_v\beta_1$ appear to recognize the same peptide sequence, i.e. Arg-Gly-Asp (RGD) on fibronectin or on vitronectin (for $\alpha_v\beta_1$).[9] Among the other binding sites identified are the peptide sequence Gly-Glu-Ala in type I collagen which is recognized by integrin $\alpha_2\beta_1$[14] and the sequence Glu-Ile-Leu-Asp-Val recognized by $\alpha_4\beta_1$ on fibronectin.[15]

Antibody and peptide blocking assays, mutational analyses and studies with integrin-deficient ("knockout") mouse embryos have

Table 4.1. Integrin complexes and their ligands

α subunit	β subunit	Ligands subunit
α_1	β_1	COL, LN
α_2		COL, LN, echo virus
α_3		COL, epiligrin, FN, LN
α_4		FN, invasin, VCAM-1
α_5		FN, denatured coll
α_6		LN
α_7		LN
α_8		Tenascin, FN, VN
α_9		?
α_v		VN, FN
α_L	β_2	ICAMs
α_M		iC3b, Factor X, Fgn, ICAMs
α_X		Fgn, iC3b
α_v	β_3	VN, FN, Fgn, Tsp, vWF, OP, denatured COL, tenascin, disintegrins, HIV Tat
α_{IIb}		VN, FN, Fgn, Tsp, vWf, COL, denatured COL, disintegrin
α_6	β_4	LN
α_v	β_5	VN, HIV Tat protein
α_v	β_6	FN, tenascin
α_4	β_7	FN, VCAM-1
α_E	β_7	E-cadherin
α_v	β_8	?

The information in this table is based on references 4, 9, 47, and 155-157. The abbreviations are: COL:collagen; FN:fibronectic; Fgn:fibrinogen; HIV:Human immunodeficiency virus; ICAM:intercellular adhesion molecule; LN:laminin; OP:osteopontin; Tat: trans activator protein of HIV; Tsp:thrombospondin; VCAM: vascular adhesion molecule; VN:vitronectin; vWF:von Willebrand factor.

shown that members of the β_1 subfamily play a critical role in a range of physiological processes. These include crucial steps during embryonic development such as the directed migration of mesodermal cells during gastrulation and the adhesion of migrating neural crest cells to ECM proteins.[5,16,18] In addition, β_1 integrins play a role in cellular differentiation both during embryogenesis and in adult tissue,[19-21] regulate matrix assembly and tissue remodeling,[22,23] play a role in inflammation and host immune responses by mediating leukocyte migration and extravasation into sites of inflammation,[24,25] and can have important effects on the growth and metastasis of cancer cells.[26-28] In many of these processes the β_1 integrins act together with other adhesion receptors of the integrin or other families in a highly coordinated manner (ref. 29, see discussion below).

THE β_2 FAMILY

The β_2 integrins are also known as leukocyte integrins. They are expressed mainly on white blood cells and mediate cell-cell interactions critical for the inflammatory process and for cell mediated immune responses.[25] The β_2 subunit can associate with one of three α chains designated, α_L, α_M and α_X.[30] Integrin $\alpha_L\beta_2$, also known as LFA-1 (lymphocyte function antigen-1), or CD11a/CD18 is expressed on lymphocytes and neutrophils and recognizes ligands ICAM-1, ICAM-2 and ICAM-3[25,31-33] on the vascular endothelium and on other cells. This integrin has been implicated in lymphocyte trafficking and leukocyte extravasation during inflammation and the host immune response. In these processes it acts as part of an adhesion cascade also involving selectins and ECM receptors of the β_1 subfamily.[25,34,35] In addition, it plays a role in stabilizing cell-cell adhesion during the antigen-specific response of T lymphocytes initiated by activation of the T cell receptor.[25,29,35,36] Integrin $\alpha_M\beta_2$ (MAC-1, CD11b/CD18) is expressed on monocytes and serves as a receptor for the C3b component of complement as well as for factor X, ICAM-1 and fibrinogen.[9,37] As is the case for the LFA-1 receptors this monocyte receptor participates in adhesive interaction required for passage through the blood vessel endothelium.[38] Receptor $\alpha_x\beta_2$ (CD11c/CD18) is expressed on neutrophils upon activation with modulators such as TNFα or phorbol esters. Similarly to $\alpha_m\beta_2$ it binds fibrinogen, and the C3b component of complement.[9,38]

As may be predicated from their specialized role in inflammation and cell-mediated immune responses, integrins of the β_2 family are normally expressed on the cell surface in an inactive state. They undergo a conformational change and become active in response to inflammatory mediators or as a consequence of a preceding interaction at the cell surface such as the cross linking of the T cell receptor[40,41] (activation of integrins as a regulatory mechanism is discussed in greater detail below).

THE β_3 FAMILY

Two integrins have been identified in this subfamily (also known as cytoadhesins), namely the platelet receptor $\alpha_{IIb}\beta_3$ (known earlier as gpIIb/IIIa)[42] and the vitronectin receptor (VNR) $\alpha_V\beta_3$.[43] Both integrins mediate adhesion to a range of matrix proteins including vitronectin, fibrinogen, fibronectin, von Willebrand factor and thrombospondin. While the RGD peptide serves as the $\alpha_v\beta_3$ ligand, $\alpha_{IIb}\beta_3$ may also recognize other structures in addition to RGD.[9] Integrin $\alpha_{IIb}\beta_3$ has been extensively studied and much of the current understanding on the regulation of integrin function is derived from these studies.[4,5,9,44,45] Similarly to integrins of the β_2 family, this receptor is expressed on resting platelets in an inactive conformational state. Upon activation by mediators such as thrombin, it undergoes the conformational alteration required for binding of soluble fibrinogen and other ligands. Binding of ligand triggers further conformational change in the receptor and is accompanied by platelet activation. This multi-level control mechanism ensures a tight regulation of the events leading to blood clotting.[45]

Under normal physiological states the expression of integrin $\alpha_v\beta_3$ is restricted. It is normally expressed on vascular endothelial cells where it is involved in attachment to, and remodeling of the subendothelial matrix, two functions which are required for neovascularization (angiogenesis) during processes such as wound healing.[46] In addition, it can be found on macrophages and osteoclasts, on smooth muscle cells, on B, T and NK cells and on chorionic villi cells of the placenta.[47] In processes such as wound healing and inflammation, its expression can be upregulated by cytokines and it mediates cell migration on extracellular matrix vitronectin.[49] Recent studies have implicated this receptor in progression of melanoma and glioblastoma where it is characteristically expressed on end-stage highly malignant cells.[48,50,51] Unlike most other α subunits, the α_v chain can associate with several different β subunits including β_1, β_5, β_6 and β_8. These heterodimers commonly recognize RGD sequences in vitronectin and/or fibronectin as well as in other matrix or plasma proteins.[7,9,47]

In addition to these major subfamilies which share not only a subunit but also common binding sites and modes of action, several other integrin heterodimers have been described. These include the α_v subfamily members alluded to previously,[9,47] the laminin receptor $\alpha_6\beta_4$ which has recently been implicated by several studies in carcinoma progression,[52,53] and integrin $\alpha_4\beta_7$ which appears to recognize the same ligands as $\alpha_4\beta_1$.[54,55]

INTEGRIN STRUCTURE AND FUNCTION

The α subunits of integrins range in size from 120 to 180 kDa and consist of a glycosylated extracellular region, a single transmembrane domain and a short cytoplasmic tail. They have been divided into two groups on the basis of structural similarities. Members of one group which includes subunits α_1, α_2 and the three subunits of the β_2 subfamily namely α_L, α_M and α_X have in common a peptide of approximately 190 amino acids designated the I domain which is inserted into the extracellular segment and appears to play a role in cation-binding and ligand recognition functions.[3,4] A second group which includes all α chains involved in recognition of the RGD sequence (α_5, $\alpha_{v\ \text{and}}$ α_{IIb}), α chains involved in laminin binding (α_3, α_6 and α_7) as well as α_8 and the PS2 α chain of *Drosophila* are cleaved post-translationally in the C-terminal half of the extracellular domain just outside the transmembrane region and the cleaved segments remain linked through disulfide bonds.[3,4] The α_4[56] and α_9[57] chains do not have either of these characteristics while α_E contains an I domain and a unique site of cleavage.[58]

Several structural motifs can be found in all α chains. These include seven repeats of a homologous segment in the extracellular domain, the N-terminal three or four of which are highly conserved and contain a consensus Asp—rich, 9 amino acid sequence which is reminiscent of divalent cation-binding domains of other proteins.[3,4,54] These domains are functionally important for the assembly, expression and function of the α/β heterodimers. In the I-domain-containing subgroup this insert replaces the first of the four cation-binding sequences both structurally and functionally.[4,9] In addition, there are at least 12 conserved cysteines in the extracellular domains of the different α subunits which are probably used to form disulfide loops.[3,4]

The β subunits (95-210 kDa) also share common structural motifs. They consist of a large glycosylated extracellular domain and a single membrane-spanning domain. With the exception of the β_4 subunit, their cytoplasmic tail is only 40-60 amino acids long and well conserved across species (e.g. 98-100% conservation for β_1). In the extracellular domain there is a high degree of conservation particularly in the positions of the cysteines. Most β subunits, with the exception of β_4,[60] β_7[61] and β_8,[62] have 56 cysteines in this domain, the majority of which are located in 4 highly conserved repeat sequences. Some of the cysteines form large disulfide-linked loops which are thought to be functionally important.[3,4,9]

Studies on several integrin complexes including mutational analyses and selective transfection with subunit-encoding cDNA revealed that following translation of the α and β subunits and N-glycosylation, the two subunits assemble in the endoplasmic reticulum to form a complex which is then transported to the Golgi apparatus for further processing including cleavage of the N-terminal domain of the α subunit.[3] Only mature complexes can be transported to the cell surface, while free, uncomplexed subunits remain in the precursor form and are eventually degraded. N-linked oligosaccharides on the integrins are usually of the complex type although high mannose structures can also be expressed. These carbohydrates are thought to be involved in integrin assembly as well as in ligand recognition and binding.[3,58,59]

Electron microscopy has shown that when expressed on the cell surface, the two subunits form a structure which has the appearance of an ovoid head composed of the N-terminal domains of both subunits and connected to the plasma membrane by two "stems" (the C-terminal portions of the extracellular domains of the subunits).[4,65] The ligand-binding site lies within the ovoid head in the interphase between the two subunits with each of the subunits contributing to it. Chemical crosslinking studies on $\alpha_V\beta_3$[66] and $\alpha_{IIb}\beta_3$[67] have placed their binding sites near the divalent cation-binding region of the α subunit (also the region of I domain insertion). In the β_1 subunits, two putative ligand-contact sites located between residues 120-182 and 220-231 have been identified.[68]

A small region spanning residues 207-218 (which is highly conserved in β_1 subunits) and a homologous region in β_3 are also critical for integrin function and may control the conformation of the adjacent ligand-binding sites.[9,69,70] In several integrins with an I domain in the α subunit, such as α_L, α_M and α_2, this domain was recently implicated both in ligand recognition and in divalent cation binding.[9]

Recent crystallography and NMR data on the RGD sequence which is recognized by several integrins provided additional insight into structural requirements for integrin-ligand binding. These studies revealed that the secondary structure of the ligand peptide is critical for recognition by the integrins.[71] This structure is determined not only by the amino acids within the RGD sequence which appear to form a loop structure,[72] but also by amino acids adjacent to this sequence[73] (for review see ref. 9).

While the extracellular domains of both integrin subunits are involved in ligand-recognition and binding and confer on the integrin receptor its ligand specificity, the cytoplasmic domains mediate post-ligand binding functions which appear to be distinct. One of the earliest integrin functions to be recognized was the ability to link with, and induce changes in, the cytoskeleton.[1] At the cellular level, integrin-mediated adhesion normally triggers rapid changes in cell shape and cell spreading. This is accompanied by the establishment of focal adhesion contacts which are transmembrane junctions between the ECM and the cytoskeleton, or of point contacts which are smaller punctuate regions of cell substratum contact where integrin subunits and cytoskeletal proteins such as talin, vinculin and α-actinin can be seen to co-localize.[4,74,75] Biochemical and molecular analyses have provided evidence that the cytoplasmic domain of the β_1 subunit associates with the talin[76] and α-actinin[77] which in turn link with actin and induce the observed changes in cell shape. This process could be inhibited in cells expressing β_1 subunits whose cytoplasmic domains were either deleted or mutated.[78] The α subunit contributes to and can determine the localization of integrins to these foci of cell-substratum contact. This is suggested by evidence that complexes with a shared β subunit but different α subunits distribute differently to the focal contacts during cells spreading. Examples include the distribution to focal contacts of $\alpha_5\beta_1$ but not $\alpha_3\beta_1$ on MRC-5 fibroblasts[79] and of $\alpha_6\beta_1$ but not $\alpha_1\beta_1$ on astrocytes.[75] Integrin $\alpha_6\beta_4$ is distinct in this respect as it localizes to hemidesmosomes in epithelial cells where it links the extracellular matrix to intermediate filaments.[80,81]

REGULATION OF INTEGRIN SYNTHESIS AND FUNCTION

Integrin-mediated cell adhesion and the consequent generation of intracellular signals can be modulated at two levels: (i) by the level of cell surface expression of a particular integrin heterodimer, and (ii) by the "activity" status of the integrin.

(i) The expression of integrin receptors is regulated at the transcriptional as well as post-transcriptional levels. The expression of the "common" subunits within the integrin subfamilies, i.e., "β_1," "α_v" and "β_2" appears to be constitutive and maintained at a high level.[82] The level of expression of a given heterodimer is therefore regulated by the rate of expression, and/or processing of the opposite "α" or "β" subunit. The level of expression of integrins can be regulated by certain cytokines and growth factors[82-85] as well as by steroid hormones[86,87] and the ECM itself.[88] Transforming growth factor β (TGFβ) appears to be a major regulator of the expression of many different integrin heterodimers. This regulation takes place at the levels of transcription and also translation.[82,83] The promoter regions of some of the genes encoding integrin subunits have been analyzed and several response elements have been identified. Interestingly, the promoter region of the $\beta3$ integrin subunit has both a vitamin D response element as well as a retinoic acid response element,[87] and the α_2 promoter has been shown to have an estrogen response element as well as an AP-2 site.[89] Although the regulation of integrin expression at the translational and post-translational levels is poorly understood, a recent report[90] indicates that the resident endoplasmic reticulum chaperone, calnexin, may be involved in regulating integrin heterodimer assembly.

(ii) Integrin function can also be regulated via changes in the affinity of the integrin for its ligand. This latter process, termed "inside-out signaling"[45] is an important aspect of integrin function. The affinity of β_1, β_2 and β_3 integrins can be altered in response to cytoplasmic signals.[91-94] In both circulating platelets and leukocytes, the binding affinities of integrins are dramatically increased by certain agonists. The mechanisms by which cytoplasmic signals regulate the integrin affinity state are not well understood, but signaling molecules such as G proteins, protein kinases, intracellular Ca^{2+} and phospholipases have been implicated.[45,95] Site-directed mutagenesis resulting in the deletion of portions of the cytoplasmic domain of both the α and β subunits has implicated various regions of the cytoplasmic domains in focal adhesion plaque localization as well as in the regulation of ligand binding affinity.[96,97] In particular, it appears that the most proximal regions to the transmembrane domain of the α subunits are involved in the regulation of ligand binding. For example, the Gly-Phe-Phe-Lys-Arg (GFFKR) motif in the cytoplasmic domain of the subunits may be particularly important since deletion of this sequence resulted in a constitutively active form of the $\alpha_{IIb}\beta3$ integrin.[44,98] Interestingly, an intracellular, Ca^{2+} binding protein, calreticulin, has been shown to bind to this motif,[99,100] and downregulation of its expression resulted in a loss of integrin-mediated cell attachment and spreading.[100] The affinity state of integrins can also be modulated by subunit-specific monoclonal antibodies which trigger the switch from a non-active to an active integrin.[101-103] The activation of integrins by such antibodies, or by phorbol esters such as PMA, which activate protein kinase C, results in an increased binding of calreticulin to the activated integrins. Disruption of calreticulin-integrin interaction by anti-calreticulin antibodies inhibits antibody as well as PMA induced integrin activation.[104] Calreticulin, therefore, by binding to the GFFKR motif, may be involved in stabilizing the high-affinity form of integrins.

The cytoplasmic domain of the β_1 subunits also regulates the affinity of integrins for their ligands. This has recently been demonstrated by overexpressing chimeric forms of the β_1 cytoplasmic domains.[105,106] The over-expressed β_1 cytoplasmic domains were found to mediate a dominant-negative effect and inhibit "inside-out" signal transduction. The most logical explanation for this finding is that the over-expressed domain competes for intracellular factors required for the regulation of the signaling process. The identity of such intracellular factors remains unknown at present.

THE ROLE OF INTEGRINS IN OUTSIDE-IN SIGNALING

The marked changes in morphology and behavior that occur when cells interact with the ECM through their integrin receptors strongly suggest that these receptors can function as transducers of extracellular signals. Differentiation of numerous cell types including keratinocytes,[107] myoblasts,[108] steosarcoma cells,[108] and neurons[84,110] is influenced by integrin function, and in some instances changes in gene expression have also been documented.[111-113] The interaction of cells with the ECM also regulates cell proliferation and cell survival since disruption of epithelial cell- or endothelial cell-matrix interactions was shown to induce apoptosis.[114,115] The findings that the onset of apoptosis in endothelial cells which are maintained in suspension can be inhibited by the incubation of these cells with anti-β_1 integrin antibodies[114] and that integrin $\alpha_v\beta_3$-mediated adhesion can rescue melanoma cells growing in 3D collagen gels from apoptosis[116] suggests that integrins transduce cell survival signals from the ECM. In fibroblasts, adhesion via integrins has also been shown to regulate the cell cycle, specifically by regulating the expression of cyclin A.[117] The fact that fibronectin- as well as integrin α_5-deficient mice fail to develop due to mesodermal and neural deficiencies[118] indicates that fibronectin-mediated signaling through integrins is critical for development.

Perhaps the strongest evidence for signal transduction via integrins comes from work carried out with T lymphocytes and monocytes. Co-stimulatory signals for the CD3-dependent proliferative response of $CD4^+$ T cells were produced by their adherence to fibronectin via $\alpha_4\beta_1$ or $\alpha_5\beta_1$ integrins, or to laminin via the $\alpha_6\beta_1$ integrin.[119,120] Similarly, engagement of the β_2 integrin, LFA-1, with I CAM-1 or with anti-LFA monoclonal antibodies demonstrates that integrins act as co-stimulators of T cell activation.[121,122] In monocytes, adhesion to various ECM proteins differentially regulates the levels and types of monocyte adherence-derived (MAD) inflammatory genes, and monoclonal antibody engagement of β_1 integrins was shown to stimulate gene expression of inflammatory mediators.[123]

Although the intracellular signaling pathways induced by integrin occupation are not fully characterized as yet, activation of the Na^+/H^+ antiporter,[124,125] changes in calcium influx and mobilization,[125] activation of the transcription factors nuclear factor κB[125] and AP-1,[124] and stimulation of tyrosine phosphorylation[126-128] all appear to be involved. Integrin-mediated tyrosine phosphorylations of paxillin,[129] pp125 focal adhesion kinase[130-132] and an as yet uncharacterized 130 kDa protein, and integrin-specific 52-55 kDa and 90 kDa proteins[133] have been reported.

The intracellular signaling pathways stimulated by integrin activation and ligation are only now beginning to unravel. These pathways are likely to be cell-type specific. Two recent reports demonstrate that in fibroblasts, cell attachment to fibronectin results in the activation of MAP kinase[134,135] which phosphorylates and activates nuclear transcription factors. This activation of MAP kinase may occur via activation of p21*ras* into a GTP-bound form. This evidence comes from two sources. Firstly, the direct antibody-mediated activation and ligation of $\alpha_2\beta_1$ integrin resulted in the accumulation of activated GTP-bound p21ras.[128] Secondly, cell attachment to fibronectin results in the association of the Grb-2/SOS[136] complex with tyrosine phosphorylated focal adhesion kinase (FAK).[135] Grb-2 binds to phosphorylated FAK via its SH2 domain and Grb-2 bound to phosphorylated tyrosine complexes activates SOS which is a p21ras-specific guanine nucleotide releasing factor.

The molecular basis for the integrin-mediated stimulation of FAK tyrosine phosphorylation remains unclear. It is likely that this phosphorylation is regulated by an upstream kinase which can physically associate with the cytoplasmic domains of integrins. Alternatively, autophosphorylation of FAK may be induced. Evidence for the existence of a kinase associated with the β_1 cytoplasmic tail has recently surfaced. A novel protein kinase termed Integrin Linked Kinase (ILK) has been identified in a yeast based two-hybrid genetic screen using the cytoplasmic tail of the β_1 subunit as bait (Hannigan and Dedhar, unpublished observations). This novel kinase may couple integrins to FAK. The initial tyrosine phosphorylation of FAK also results in the binding of C-SRC to FAK via its SH2 domain.[135] This binding is thought to activate the kinase activity of C-SRC resulting in the phosphorylation of additional tyrosine residues on FAK,[137] one of them being the tyrosine for Grb-2 binding.

In monocytes, which lack FAK expression, integrin ligation has been shown to activate the transcription factor, NFκB.[125] An intermediate of this activation seems to be a 76 kDa protein which becomes tyrosine phosphorylated by integrin ligation.[138]

The actin cytoskeleton also appears to play a central role in integrin-mediated signal transduction. Thus the phosphorylation of FAK and activation of MAP kinase require the existence of an organized actin cytoskeleton.[126,134] The small GTP-binding protein Rho has been shown to regulate actin cytoskeleton reorganization,[139] as well as integrin-mediated adhesion.[140] Although the regulation of Rho activity remains unclear, it may be regulated by receptor-coupled G-proteins since lysophosphatidic acid (LPA) and Bombesin, both of which bind to receptors coupled to G-proteins, induce a rapid actin cytoskeleton reorganization and also stimulate FAK in this signaling pathway.[141] Recent evidence suggests that this may indeed be the case since the inhibition of Rho activity by ADP ribosyl transferase also results in the inhibition of FAK phosphorylation.[142]

The findings that some of the signaling pathways activated by integrins and growth factor receptors are similar suggest the possibility of cross-talk between ECM induced and hormone, growth factor or cytokine induced signaling. Indeed some of the earliest studies in this area suggested the existence of such cross-talk, since interleukin-1 induced osteoblastic differentiation of osteosarcoma cells was shown to require integrin function,[143] fibronectin binding to $\alpha_5\beta_1$ was shown to induce the AP-1 transcription factor also required for IL-2 transcription[144] and $\alpha_v\beta_3$ was identified as a co-stimulator for IL-4 production.[145] More recently it has been found that insulin-mediated stimulation of tyrosine phosphorylation of the insulin receptor substrate, IRS-1 results in the association of phosphorylated IRS-1 with the integrin $\alpha_v\beta_3$. Indeed insulin induced DNA synthesis is enhanced when cells are plated on vitronectin, the ligand for $\alpha_v\beta_3$.[146]

The cytoplasmic domains of integrin subunits contain a highly conserved amino acid sequence, KXGFFKR, which has been shown to be the binding site for calreticulin.[99,100] Surprisingly, an almost identical sequence is present in the DNA binding domains of the nuclear hormone receptor transcription factors, and calreticulin has been shown to bind to this sequence.[109,147] Calreticulin binding to hormone receptors results in the inhibition of receptor-mediated transcription and therefore this protein behaves as a negative regulator of transcription. Since ligand-bound high-affinity forms of integrins bind calreticulin, such integrins may sequester calreticulin to the plasma membrane-cytoplasm interface allowing for fully active transcription via the nuclear hormone receptors. Thus, the binding of calreticulin to integrins may serve a dual function: that of stabilizing the integrins in the high-affinity conformational state, as discussed in the previous section, and secondly of sequestering calreticulin away from the nucleus, thus enhancing gene transcription.

These types of data clearly demonstrate that the activity of hormones, growth factors and cytokines are modulated by cell-ECM interactions via integrins, and the dissection of the molecular basis for such cross-talk provides an interesting challenge for the future. Possible signaling pathways activated by integrins are depicted in Figure 4.1. Although some of these are hypothetical at the present time, many of the molecular players are rapidly being identified.

THE ROLE OF INTEGRINS IN ADHESION CASCADES

In addition to growth factors and hormones, integrin function is also coordinated with the activity of members of the other major families of adhesion receptors. As already indicated, some members of the Ig supergene family such as the ICAMs and VCAM-1 actually serve as ligands for integrin receptors LFA-1 and $\alpha_4\beta_1$ respectively.[10,25,32] In addition, integrins have been implicated in several adhesion cascades which involve one or several other types of adhesion receptors. A well studied example of this cooperative interaction between diverse cell adhesion mechanisms can be found in the multistep interaction between leukocytes and the vascular endothelium during leukocyte recruitment into sites of inflammation.[25] During this process, selectin expression on the endothelial cells is induced by inflammatory mediators such as IL-1 and TNFα. The selectins (E and P) are in turn recognized by carbohydrate ligands expressed on the leukocytes (see chapter 6) and under flow conditions this recognition results in successive leukocyte attachment to, and detachment from the endothelium seen cumulatively as leukocyte rolling.[148] This intermittent contact with the endothelial cells is the signal required for activation of leukocyte integrins $\alpha_4\beta_1$ and LFA-1 and these, in turn, mediate high affinity, firm adhesion to the endothelium via their counter-receptors VCAM-1 and ICAMs whose expression is also upregulated at sites of inflammation.[149] This adhesion is a prerequisite for leukocyte extravasation into the subendothelial space and migration into the inflamed tissue, processes which involve recognition of, and adhesion to ECM proteins via the appropriate integrin receptors.[29]

Similar cooperation between different classes of receptors is also likely to occur during the process of cancer metastasis which involves multiple interactions between the disseminating cancer cells and the vascular endothelium, both during intravasation and extravasation of the malignant cells.[150,151] In recent studies we found

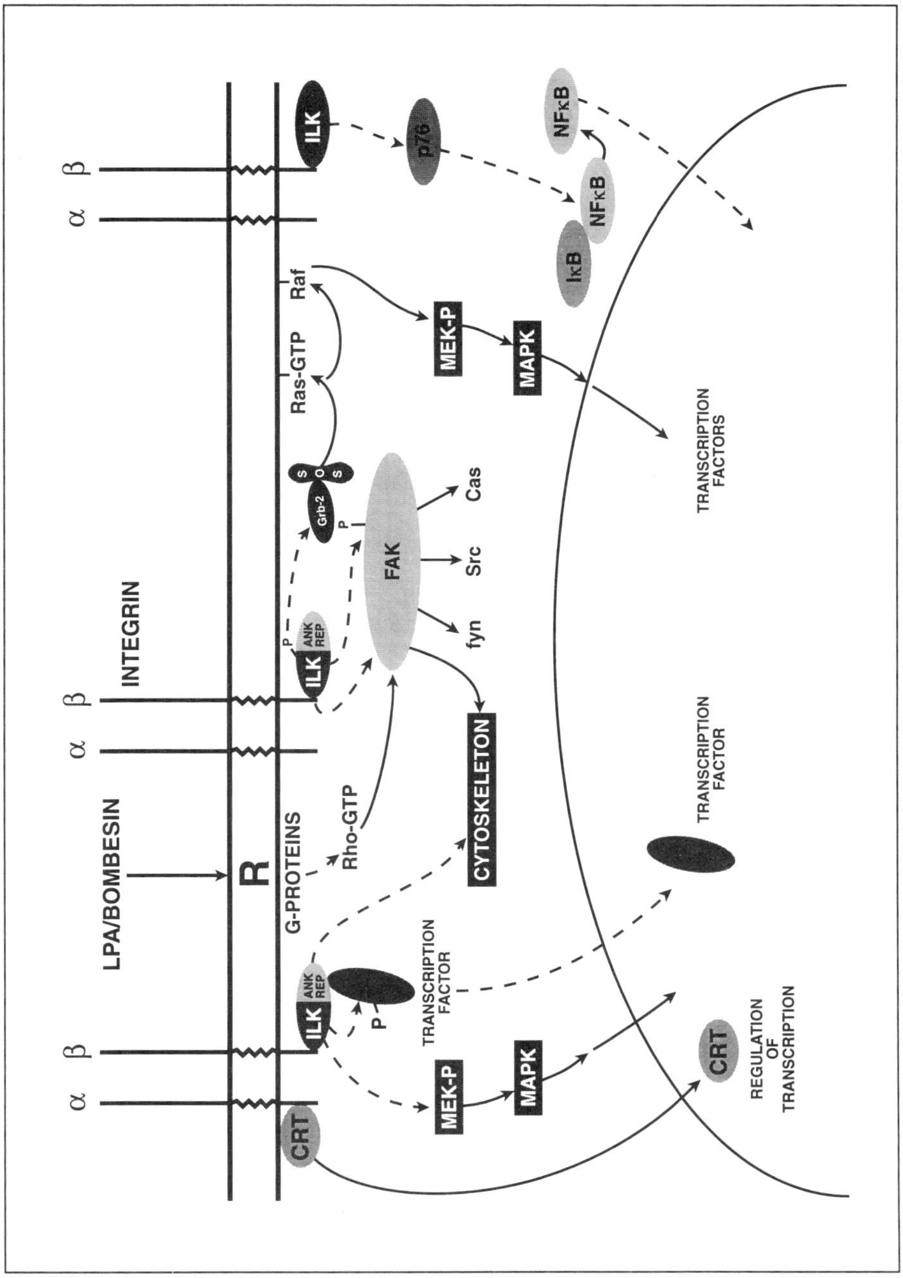

Fig. 4.1. A summary of currently established (solid lines) and hypothesized (broken lines) signal transduction pathways involving integrins. FAK: Focal Adhesion Kinase; ILK: Integrin Linked Kinase; CRT: Calreticulin; SOS: Son of Sevenless—a Drosophila homologue of ras guanine nucleotide releasing factor; LPA: lysophosphatidic acid.

that receptor E selectin mediates the adhesion of highly metastatic carcinoma H-59 cells[152] to the hepatic sinusoidal endothelium, an interaction required for the establishment of hepatic metastases. In vitro adhesion assays suggested that this interaction is part of an adhesion cascade also involving integrins of the β_1 subfamily which are expressed on the tumor cells (Brodt et al manuscript submitted). Similarly, Lafrenie et al recently described the cooperative involvement of integrins $\alpha_4\beta_1$ and $\alpha_v\beta_3$ with E selectin in adhesion of human cancer cells to IL-1 activated vascular endothelial cells.[153]

Other examples of cooperation between integrins and other adhesion receptors can be found in the multistep processes of platelet aggregation where activation of the constitutively expressed $\alpha_{II}\beta_3$ integrin occurs only in response to binding of agonists such as thrombin or collagen to their respective cell surface receptors,[42] in lymphocyte homing to secondary lymphoid organs which requires the coordinated action of at least three types of receptors namely, L selectin, integrins LFA-1 and $\alpha_4\beta_P$ and the widely expressed receptor CD44,[35,149,154] and in T lymphocyte activation which involves the cooperative action of the T cell receptor recognizing antigen in the context of the MHC complex on one hand and an LFA-1/ICAM recognition on the other.[4,36,102,121]

Taken together, the evidence clearly demonstrates that the integrins often act as a link in a chain of adhesive interactions between neighboring cells or between cells and their extracellular microenvironment. In this fashion integrin-mediated recognition becomes both dependent on, and a prerequisite for, the function and activity of other adhesion and signal transducing systems. As is discussed in detail in chapter 11, the process of malignant transformation frequently results in loss of normal integrin expression and function or in the deregulated expression and distribution of new integrins. In view of the essential and inter-linked role that these receptors play in the regulation of cellular communication, it is not surprising that such disruption in integrin function can be the cause and/or manifestation of profound alterations in the normal patterns of cellular behavior.[5,28]

REFERENCES

1. Hynes RO. Integrins: a family of cell surface receptors. Cell 1987; 48:549-54.
2. Schwartz MA. Transmembrane signalling by integrins. Trends Cell Biol 1992; 2:304-7.
3. Hemler ME. Structures and functions of VLA proteins and related integrins. In: McDonald JA, Mecham RP, ed. Receptors for Extracellular Matrix. San Diego: Academic Press 1991; 255-300.
4. Hynes RO. Integrins: versatility, modulation and signaling in cell adhesion. Cell 1992; 69:11-25.
5. Hynes RO, Landers AD. Contact and adhesive specificities in the associations, migrations and targeting of cells and axons. Cell 1992; 68:303-22.
6. Damsky CH, Werb Z. Signal transduction by integrin receptors for extracellular matrix: cooperative processing of extracellular information. Curr Opin Cell Biol 1992; 4:772-81.
7. Cheresh DA. Structural and biologic properties of integrin-mediated cell adhesion. Clin Lab Med 1992; 12:217-36.
8. Akiyama SK, Nagata K, Yamada KM. Cell surface receptors for extracellular matrix components. Biochim Biophys Acta 1990; 1031:91-110.
9. Haas TA, Plow EF. Integrin-ligand interactions: a year in review. Curr Opin Cell Biol 1994; 6:656-62.
10. Elices MJ, Osborn L, Takada Y et al. VCAM-1 on activated endothelium interacts with the leukocyte integrin VLA-4 at a site distinct from VLA-4 fibronectin binding site. Cell 1990; 60:577-84.
11. Pepinsky B, Hession C, Chen LL et al. Structure/function studies on vascular cell adhesion molecule-1. J Biol Chem 1992; 267:17820-6.
12. Weitzman JB, Pasqualini R, Takada Y et al. The function and distinctive regulation of the integrin VLA-3 in cell adhesion, spreading, and homotypic cell aggregation. J Biol Chem 1993; 268:8651-7.
13. Symington BE, Takada Y, Carter WG. Interaction of integrins $\alpha_3\beta_1$ and $\alpha_2\beta_1$: potential role in keratinocyte intercellular adhesion. J Cell Biol 1993; 120:523-35.
14. Staatz WD, Fok KF, Zutter MM et al. Identification of a tetrapeptide recognition sequence for the $\alpha_2\beta_1$ integrin in collagen. J Biol Chem 1991; 266:7363-7.
15. Wayner EA, Garcia PA, Humphries MJ et al. Identification and characterization of the T lymphocyte adhesion receptor for an alternative cell attachment domain (CS-1) in plasma fibronectin. J Cell Biol 1989; 109:1321-30.
16. Venstrom KA, Reichardt LF. Extracellular matrix. Role of extracellular matrix molecules and their receptors in the nervous system. FASEB 1993; 7:996-1003.
17. Yang JT, Rayburn H, Hynes RO. Embryonic mesodermal defects in α_5-integrin-deficient mice. Development 1993; 119:1093-105.

18. Hynes RO. Genetic analyses of cell-matrix interactions in development. Curr Opin Genet Develop 1994; 4:569-74.
19. Hotchin NA, Kovach NL, Watt FM. Functional down-regulation of $\alpha_5\beta_1$ integrin in keratinocytes is reversible but commitment to terminal differentiation is not. J Cell Sci 1993; 106:1131-8.
20. Damsky C, Sutherland A, Fisher S. Extracellular matrix5: Adhesive interactions in early mammalian embryogenesis, implantation, and placentation. FASEB 1993; 7:1320-9.
21. Streuli CH, Bailey N, Bissell MF. Control of mammary epithelial differentiation: basement membrane induces tissue-specific gene expression in the absence of cell-cell interaction and morphological polarity. J Cell Biol 1991; 115:1383-95.
22. Rusolahti E. Integrins. J Clin Invest 1991; 87:1-5.
23. Damsky C, Tremble P, Werb Z. Signal transduction via the fibronectin receptor: do integrins regulate matrix remodeling? Matrix (Suppl) 1992; 1:184-91.
24. Sanchez-Mateos P, Arroyo AG, Balboa MA et al. Post-receptor occupancy events in leukocytes during β_1 integrin-ligand interactions. Eur J Immunol 1993; 23:2642-8.
25. Springer TA. Traffic signals for lymphocyte recirculation and leukocyte emigration: the multistep paradigm. Cell 1994; 76:301-14.
26. Juliano RL. The role of β_1 integrins in tumors. Sem Cancer Biol 1993; 4:277-83.
27. Chan BM, Matsuura N, Takada Y et al. In vitro and in vivo consequences of VLA-2 expression on rhabdomyosarcoma cells. Science 1991; 251:1600-2.
28. Schwartz MA. Signaling by integrins: implications for tumorigenesis. Cancer Res 1993; 53:1503-6.
29. Hynes RO. The impact of molecular biology on models for cell adhesion. BioEssays 1994; 16:663-9.
30. Mazzone A, Ricevuti G. Leukocyte CD11/CD18 integrins—biological and clinical relevance. Haematologica 1995; 80:161-75.
31. Sligh JE, Ballantyne CM, Rich SS et al. Inflammatory and immune responses are impaired in mice deficient in intercellular adhesion molecule-1. Proc Natl Acad Sci USA 1993; 90:8529-33.
32. Campanero MR, Delpozo MA, Arroyo AG et al. ICAM-3 interacts with LFA-1 and regulates the LFA-1/ICAM-1 cell adhesion pathway. J Cell Biol 1993; 123:1007-16.
33. Li R, Nortamo P, Valmu L et al. A peptide from ICAM-2 binds to the leukocyte integrin CD11A/CD18 and inhibits endothelial cell adhesion. J Biol Chem 1993; 268:17513-8.
34. Mackay CR and Imhof BE. Cell adhesion in the immune system. Immunol Today 1993:14:99-102.
35. Picker LJ, Treer JR, Nguyen M et al. Coordinate expression of β_1 and β_2 integrin "activation" epitopes during T cell responses in secondary lymphoid tissue. Eur J Immunol 1993; 23:2751-7.

36. Springer TA. Adhesion receptors of the immune system. Nature 1990; 346:425-34.
37. Wright SD. Adhesion-promoting receptors on phagocytes. Agents & Actions—Supplements 1991; 35:23-6.
38. Diamond MS, Staunton DE, Marlin SD et al. Binding of the integrin Mac-1 (CD11b/CD18) to the third immunoglobulin-like domain of ICAM-1 (CD54) and its regulation by glycosylation. Cell 1991; 65:961-71.
39. Loike JD, Sodeik B, Cao L et al. CD11c/CD18 on neutrophils recognizes a domain at the N terminus of the A α chain of fibrinogen. Proc Natl Acad Sci USA 1991; 88:1044-8.
40. van Kooyk Y, Weder P, Hogervorst F et al. Activation of LFA-1 through a Ca^{2+}-dependent epitope stimulates lymphocyte adhesion. J Cell Biol 1991; 112:345-54.
41. Hermanowski-Vosatka A, Van Strijp JAG, Swiggard WJ et al. Integrin modulating factor-1: a lipid that alters the function of leukocyte integrins. Cell 1992; 68:341-52.
42. Ginsberg MH, Du XP, O'Toole TE et al. Platelet integrins. Tromb and Haemost 1993; 70:87-93.
43. Cheresh DA. Integrins in trombosis, wound healing and cancer. Biochem Soc Trans 1991; 19:853-8.
44. O'Toole TE, Madekman D, Forsyth J et al. Modulation of affinity of integrin $\alpha_{IIb}\beta_3$ (GPIIb-IIIa) by the cytoplasmic domain of α_{IIb}. Science 1991; 254:845-7.
45. Ginsberg MH, Du X and Plow EF. Inside-out integrin signalling. Curr Opin Cell Biol 1992; 4:766-71.
46. Brooks PC, Clark RAF, Cheresh DA. Requirement of vascular integrin $\alpha_v\beta_3$ for angiogenesis. Science 1994; 264:569-71.
47. Felding-Habermann B, Cheresh DA. Vitronectin and its receptors. Curr Opin Cell Biol 1993; 5:864-8.
48. Nip J, Brodt P. The role of the integrin vitronectin receptor in melanoma metastasis. Cancer Metastasis Rev; (in press).
49. Leavesley DI, Schwartz MA, Rosenfeld M et al. Integrin β_1- and β_3-mediated endothelial cell migration is triggered through distinct signaling mechanisms. J Cell Biol 1993; 121:163-70.
50. Albelda SM, Mette SA, Elder DE et al. Integrin distribution in malignant melanoma: association of the β_3 subunit with tumor progression. Cancer Res 1990; 50:6757-64.
51. Sanders LC, Felding-Habermann B, Mueller BM et al. Role of α_v integrins and vitronectin in human melanoma cell growth. Cold Spring Harbor Symposia on Quantitative Biology 1992; (LVII) :233-40.
52. Lee EC, Lotz MM, Steele GD Jr et al. The integrin $\alpha_6\beta_4$ is a laminin receptor. J Cell Biol 1992; 117:671-8.
53. Tennenbaum T, Weiner AK, Belanger AJ et al. The suprabasal expression of $\alpha_6\beta_4$ integrin is associated with a high risk for malig-

nant progression in mouse skin carcinogenesis. Cancer Res 1993; 53:4803-10.

54. Ruegg C, Postigo AA, Sikorski EE et al. Role of integrin $\alpha_4\beta_7/\alpha_4\beta_P$ in lymphocyte adherence to fibronectin and VCAM-1 and in homotypic cell clustering. J Cell Biol 1992; 117:179-89.
55. Lazorovits AI, Karsh J. Differential expression in rheumatoid synovium and synovial fluid of $\alpha_4\beta_7$ integrin—a novel receptor for fibronectin and vascular cell adhesion molecule-1. J Immunol 1993; 151:6482-9.
56. Teixidó J, Parker CM, Kassner PD et al. Functional and structural analysis of VLA-4 integrin α_4 subunit cleavage. J Biol 1992; 267:1786-91.
57. Palmer EL, Ruegg C, Gerrando R et al. Sequence and tissue distribution of the integrin α_9 subunit, a novel partner of β_1 that is widely distributed in epithelia and muscle. J Cell Biol 1994; 124:395.
58. Shaw SK, Cepek KL, Murphy EA et al. Molecular cloning of the human mucosal lymphocyte integrin α_E subunit. Unusual structure and restricted RNA distribution. J Biol Chem 1994; 269:6016-25.
59. Tuckwell DS, Brass A, Humphries MJ. Homology modelling of integrin EF-hands: evidence for widespread use of a conserved cation-binding site. Biochem J 1992; 285:325-31.
60. Suzuki S, Naitoh Y. Amino acid sequence of a novel integrin β_4 subunit and primary expression of the mRNA in epithelial cells. EMBO J 1990; 9:757-63.
61. Yuan QA, Jiang WM, Krissansen GW et al. Cloning and sequence analysis of a novel β_2-related integrin transcript from T lymphocytes: homology of integrin cysteine-rich repeats to domain III of laminin B chains. Int Immunol 1990; 2:1097-108.
62. Moyle M, Napier MA, McLean JW. Cloning and expression of a divergent integrin subunit β_8. J Biol Chem 1991; 266:19650-8.
63. Spiro RC, Laufer DM, Perry SK et al. Effect of inhibitors of N-linked oligosaccharide processing on the cell surface expression of a melanoma integrin. J Cell Biochem 1989; 41:37-45.
64. Akiyama SK, Yamada SS, Yamada KM. Analysis of the role of glycosylation of the human fibronectin receptor. J Biol Chem 1989; 264:18011-18.
65. Nermut MV, Green NM, Eason P et al. Electron microscopy and structural model of human fibronectin receptor. EMBO J 1988; 7:4093-9.
66. Smith JW, Cheresh DA. Integrin ($\alpha_v\beta_3$)-ligand interaction. J Biol Chem 1990; 265:2168-72.
67. D'Souza SE, Ginsberg MH, Burke TA et al. Localization of an Arg-Gly-Asp recognition site within an integrin adhesion receptor. Science 1988; 242:91-3.
68. Shih DT, Edelman JM, Horowitz AF et al. Structure/function analysis of the integrin β_1 subunit by epitope mapping. J Cell Biol 1993; 122:1361-71.

69. Takada Y, Puzon W. Identification of a regulatory region of integrin β_1 subunit using activating and inhibiting antibodies. J Biol Chem 1993; 268:17597-601.
70. Steiner B, Trzeciak A, Plenninger G et al. Peptides derived from a sequence within β_3 integrin bind to platelet $\alpha_{IIb}\beta_3$ (GPIIbIIIa) and inhibit ligand binding. J Biol Chem 1993; 268:6870-3.
71. Johnson WC Jr, Pagano TG, Basson CT et al. Biologically active Arg-Gly-Asp oligopeptides assume a type II β-turn in solution. Biochemistry 1993; 32:268-73.
72. Dickinson CD, Veerapandian B, Dai XP et al. Crystal structure of the tenth type III cell adhesion module of human fibronectin. J Mol Biol 1994; 236:1079-92.
73. Pfaff M, Aumailley M, Specks U et al. Integrin and Arg-Gly-Asp dependence of cell adhesion to the native and unfolded triple helix of collagen type VI. Exp Cell Res 1993; 206:167-76.
74. Duband JL, Rocher S, Chen WT et al. Cell adhesion and migration in the early vertebrate embryo: location and possible role of the putative fibronectin receptor complex. J Cell Biol 1986; 102:160-78.
75. Tawil N, Wilson P, Carbonetto S. Integrins in point contacts mediate cell spreading: factors that regulate integrin accumulation in point contacts vs focal contacts. J Cell Biol 1993; 120:261-71.
76. Horwitz A, Duggan E, Buck C et al. Interaction of plasma membrane fibronectin receptor with talin—a transmembrane linkage. Nature 1986; 320:531-3.
77. Otey CA, Pavalko FM, Burridge K. An interaction between α-actinin and the β_1 integrin subunit in vitro. J Cell Biol 1990; 111:721-9.
78. Marcantonio EE, Guan JL, Trevithick JE et al. Mapping of the functional determinants of the integrin β^1 cytoplasmic domain by site-directed mutagenesis. Cell Regul 1991; 1:596-604.
79. Elices MJ, Urry LA, Hemler ME. Receptor functions for the integrin VLA-3: fibronectin, collagen, and laminin binding are differentially influenced by Arg-Gly-Asp peptide and by divalent cations. J Cell Biol 1991; 112:169-81.
80. Sonnenberg A, Calafat J, Janssen H et al. Integrin α6/β4 complex is located in hemidesmosomes, suggesting a major role in epidermal cell-basement membrane adhesion. J Cell Biol 1991; 113:907-17.
81. Tozeren A, Kleinman HK, Wu S et al. Integrin $\alpha_6\beta_4$ mediates dynamic interactions with laminin. J Cell Sci 1994; 1:3153-63.
82. Heino J, Ignotz RA, Hemler ME et al. Regulation of cell adhesion receptors by transforming growth factor β. Concomitant regulation of integrins that share a common β subunit. J Biol Chem 1989; 264:380-8.
83. Dedhar S. Regulation of expression of the cell adhesion receptors, integrins by recombinant human interluekin-1 in human osteosarcoma cells. Inhibition of cell proliferation and stimulation of alkaline phosphatase activity. J Cell Physiol 1989; 138:291-9.

84. Dedhar S, Gray V, Roberston K et al. Differential regulation of expression of specific integrin receptors by nerve growth factor and transforming growth factor β during differentiation of human neuroblastoma cells. Mol Cell Diff 1993; 1:1-20.
85. Chen JD, Kim JP, Zhang K et al. Epidermal growth factor (EGF) promotes human keratinocyte locomotion on collagen by increasing the α_2 integrin subunit. Exp Cell Res 1993; 209:216-23.
86. Dedhar S, Robertson K, Gray V. Induction of expression of the $\alpha_v\beta_1$ and $\alpha_v\beta_3$ integrin heterodimers during retionic acid-induced neuronal differentiation of murine embryonal carcinoma cells. J Biol Chem 1991; 266:21846-52.
87. Xu C, Ross FP, Zhang L et al. Cloning of the promoter for the avian integrin β_3 subunit gene and its regulation by 1,25 dihydroxyvitamin D3. J Biol Chem 1993; 268:27371-80.
88. Kruk PA, Vitto VS, Firth JD et al. Reciprocal interactions between human ovarian surface epithelial cells and adjacent extracellular matrix. Exp Cell Res 1994; 215:97-108.
89. Zutter MM, Santoro SA, Painter AS et al. The human α_2 integrin gene promoter: identification of positive and negative regulatory elements important for cell type and developmentally restricted gene expression. J Biol Chem 1994; 269:463-9.
90. Lenter M, Vestweber D. The integrin chains β_1 and α_6 associate with the chaperone calnexin prior to integrin assembly. J Biol Chem 1994; 269:12263-8.
91. Bennett JS, Vilaire G. Exposure of platelet fibrinogen receptors by ADP and epinephrine. J Clin Invest 1979; 64:1393-1401.
92. Marguerie CA, Plow EF, Eddington TS. Human platelets possess an inducible and saturable receptor specific for fibrinogen. J Biol Chem 1979; 254:5357-63.
93. Altieri DC, Bader R, Mannucci PM et al. Oligospecificity of the cellular adhesion receptor Mc-1 encompasses an inducible recognition specificity for fibrinogen. J Cell Biol 1988; 107:1893-1900.
94. Faull RJ, Kovach NL, Harlan JM et al. Stimulation of integrin-mediated adhesion of T lymphocytes and monocytes: two mechanisms with divergent biological consequences. J Exp Med 1995; 179:1307-16.
95. Shattil SJ, Cunningham M, Wiedmar T et al. Regulation of glycoprotein IIb-IIIa receptor function studied with platelets permeabilized by the pore-forming complement protein C5b-9. J Biol Chem 1992; 267:18424-31.
96. Bauer JS, Varner J, Scheiner C et al. Functional role of the cytoplasmic domain of the integrin α_5 subunit. J Cell Biol 1993; 122:209-21.
97. Kawaguchi S, Bevgelson JM, Finberg RW et al. Integrin α_2 cytoplasmic domain deletion effects: loss of adhesive activity parallels ligand-independent recruitment into focal adhesions. Mol Biol Cell 1994; 5:977-88.

98. Williams MJ, Hughes PE, O'Toole TE et al. The inner world of cell adhesion: integrin cytoplasmic domains. Trends Cell Biol 1994; 4:109-12.
99. Rojiani M, Finlay BB, Gray V et al. In vitro interaction of a polypeptide homologous to human Ro/SS-A antigen (Calreticulin) with a highly conserved amino acid sequence in the cytoplasmic domain of integrin α subunits. Biochemistry 1991; 30:9859-66.
100. Leung-Hagesteijn C, Milankov K, Michalak M et al. Cell attachment to extracellular matrix substrates is inhibited upon downregulation of expression of calreticulin, an intracelular integrin α-subunit-binding protein. J Cell Sci 1994; 107:589-600.
101. Stupack D, Shen C, Wilkins JA. Control of lymphocyte integrin function: evidence for multiple contributing factors. Cell Immunol 1994; 155:237-80.
102. Shimizu Y, van Seventer G, Horgan KJ et al. Regulated expressionand binding of three VLA(β_1) integrin receptors on T cells. Nature 1990; 354:250-3.
103. Van Kooyk Y, van de Wiel-van Kamenade P, Weder P et al. Enhancement of LFA-1-mediated cell adhesion by triggering through CD2 or CD3 on T lymphocytes. Nature 1989; 342:811-3.
104. Coppolino M, Leung-Hagesteijn C, Dedhar S et al. Inducible interaction of integrin $\alpha_2\beta_1$ with calreticulin: Dependence on the activation-state of the integrin. J Biol Chem 1995; (in press).
105. Chen X, O'Toole TE, Shipley T et al. "Inside-out" signal transduction inhibited by isolated integrin cytoplasmic domains. J Biol Chem 1994; 269:18307-10.
106. Lukashev ME, Sheppard D, Pytela R. Disruption of integrin function and induction of tyrosine phosphorylation by the autonomously expressed β_1 integrin cytoplasmic domain. J Biol Chem 1994; 269:18311-4.
107. Adams J, Watt F. Fibronectin inhibits the terminal differentiation of human keratinocytes. Nature 1989; 340:307-9.
108. Menko AS, Boettiger D. Occupation of the extracellular matrix receptor, integrin, is a control point for myogenic differentiation. Cell 1987; 51:51-7.
109. Dedhar S, Rennie PS, Shago M et al. Inhibition of nuclear hormone receptor activity by calreticulin. Nature 1994; 367:480-3.
110. Rossino P, Gavazzi I, Timpl R et al. Nerve growth factor induces increased expression of a laminin-binding integrin in rat pheochromocytoma PC12 cells. Exp Cell Res 1990; 189:100-8.
111. Dike LE, Farmer S. Cell adhesion induces expression of growth-associated genes in suspension-arrested fibroblasts. Proc Natl Acad Sci USA 1988; 85:6792-6.
112. Werb Z, Tremble PM, Behrendtsen O et al. Signal transduction through the fibronectin receptor induces collagenase and stromeysin gene expression. J Cell Biol 1989; 109:877-89.

113. Streuli CH, Schmidhauser C, Kobrin M et al. Extracellular matrix regulates expression of the TGFβ-1 gene. J Cell Biol 1993; 120:253-60.
114. Meredith J, Bazeli B, Schwartz, M. The extracellular matrix as a survival factor. Mol Biol Cell 1993; 4:953-61.
115. Frisch S, Francis H. Disruption of epithelial cell-matrix interactions induces apoptosis. J Cell Biol 1994; 124:619-26.
116. Montgomery AMP, Reisfeld RA, Cheresh DA. Integrin $\alpha_v\beta_3$ rescues melanoma cells from apoptosis in three-dimensional dermal collagen. Proc Natl Acad Sci USA 1994; 91:8856-60.
117. Guadagno TM, Ohtsubo M, Roberts JM et al. A link between cyclin A expression and adhesion-dependent cell cycle progression. Science 1993; 262:81572-5.
118. George EL, Georges-Labouerse EN, Patel-King RS et al. Defects in mesoderm, neural tube and vascular development in mouse embryos lacking fibronectin. Development 1993; 119:1079-91.
119. Shimizu Y, van Seventer G, Horgan K et al. Costimulation of proliferative responses of resting $CD4^+$ T cells by interaction of VLA4 and VLA5 with fibronectin or VLA6 with laminin. J Immunol 1990; 145:59-67.
120. Nojima Y, Humphries M, Mould A et al. VLA-4 mediates CD3-dependent CD4+ T-cells activation via the CSI alternatively spliced domain of fibronectin. J Exp Med 1990; 172:1185-92.
121. van Seventer G, Newman W, Shimizu Y et al. Analysis of T cell stimulation by superantigen plus major histocompatibility complex class II molecules or by CD3 monoclonal antibody: costimulation by purified adhesion ligands VCAM-1, ICAM-1, but not ELAM-1. J Exp Med 1991; 174:901-13.
122. van Seventer G, Shimizu Y, Shaw S. Roles of multiple accessory molecules in T-cell activation. Curr Opin Immunol 1991; 3:294-303.
123. Yurochko A, Liu D, Eierman D et al. Integrins as a primary signal transduction molecule regulating monocyte immediate-early gene induction. Proc Natl Acad Sci USA 1992; 89:9034-8.
124. Schwartz MA, Lechene C, Ingber DE. Insoluble fibronectin activates the Na^+/H^+ antiporter by clustering and immobilizing integrin $\alpha_5\beta_1$, independent of cell shape. Proc Natl Acad Sci USA 1991; 88:7849-53.
125. Juliano R, Haskill S. Signal transduction from the extracellular matrix. J Cell Biol 1993; 120:577-85.
126. Guan J, Trevethick J, Hynes R. Fibronectin/integrin interaction induces tyrosine phosphorylation of a 120 kDa protein. Cell Regul 1991; 2:951-64.
127. Kornberg L, Earp H, Turner C et al. Signal transduction by integrins: Increased protein tyrosine phosphorylation caused by clustering of β_1 integrins. Proc Natl Acad Sci USA 1991; 88:8392-6.

128. Kapron-Bras C, Fitz-Gibbon L, Jeevaratnam P et al. Stimulation of tyrosine phosphorylation and accumulation of GTP-bound $p21^{ras}$ upon antibody-mediated $\alpha_2\beta_1$ integrin activation in T-lymphoblastic cells. J Biol Chem 1993; 268:20701-4.
129. Burridge K, Turner C, Romer L. Tyrosine phosphorylation of paxillin and pp125 FAK accompanies cell adhesion to extracellular matrix: a role in cytoskeletal assembly. J Cell Biol 1992; 119:893-904.
130. Schaller MD, Borgman CA, Cobb BS et al. $pp125^{FAK}$, a structurally distinctive protein-tyrosine kinase associated with focal adhesions. Proc Natl Acad Sci USA 1992; 89:5192-6.
131. Kornberg L, Earp H, Parsons T et al. Cell adhesion or integrin clustering increases phosphorylation of a focal adhesion associated tyrosine kinase. J Biol Chem 1992; 267:23439-44.
132. Guan JL, Shalloway D. Regulation of focal adhesion associated protein tyrosine kinase by both cellular adhesion and oncogenic transformation. Nature 1992; 358:390-2.
133. Jewell K, Kapron-Bras C, Jeevaratnam P et al. Stimulation of tyrosine phosphorylation of distinct proteins in response to antibody-mediated ligation and clustering of α_3 and α_6 integrins. J Cell Sci 1995; 108:1165-74.
134. Chen Q, Kinch MS, Lin TH et al. Integrin-mediated cell adhesion activates mitogen-activated protein kinases. J Biol Chem 1994; 269:26602-5.
135. Schlaepjer DD, Hanks SK, Hunter T et al. Integrin-mediated signal transduction linked to *ras* pathway by GRB-2 binding to focal adhesion kinase. Nature 1994; 372:786-91.
136. Rozakis-Adcock M, McGlade J, Mbanolu G et al. Association of the Shc and Grb2/sem5 SH2 containing proteins is implicated in activation of the Ras pathway by tyrosine kinases. Nature 1992; 360:689-92.
137. Calalb MB, Polte TR, Hanks SK. Tyrosine phosphorylation of FAK at sites within the catalytic domain regulates kinase activity: a role for Src-family kinases. Mol Cell Biol 1995; 15:954-63.
138. Lin TH, Ywochko A, Kornberg L et al. The role of protein tyrosine phosphorylation in integrin-mediated gene induction in monocytes. J Cell Biol 1994; 126:1585-93.
139. Ridley AJ, Hall A. The small GTP-binding protein rho regulates the asembly of focal adhesions and actin stress fibers in response to growth factors. Cell 1992; 70:389-99.
140. McGlade J, Brunkhorst B, Anderson D et al. The amino terminal region of GAP regulates cytoskeletal structure and cell adhesion. EMBO J 1993; 12:3073-81.
141. Zachary I, Sinnet-Smith J, Rozengunt E. Bombesin, vasopressin, and endothelin stimulation of tyrosine phosphorylation in Swiss 3T3 cells. Identification of a novel tyrosine kinase as a major substrate. J Biol Chem 1992; 267:19031-4.

142. Kumagai N, Morii N, Fugisawa K et al. ADP-ribosylation of Rho p21 inhibits lysophosphatidic acid-induced protein tyrosine phosphorylation and phosphatidylinositol 3-kinase activation in Swiss 3T3 cells. J Biol Chem 1993; 268:24535-8.
143. Dedhar S. Signal transduction via the β_1 integrins is a required intermediate in interleukin-1 induced phenotypic differentiation of human osteosarcoma cells. Exp Cell Res 1989; 183:207-14.
144. Yamada A, Nikaido T, Nojima Y et al. Activation of human CD4 T lymphocytes. Interaction of fibronectin with VLA-5 receptor on CD4 cells induces the AP-1 transcription factor. J Immunol 1991; 146:53-6.
145. Roberts K, Yokoyama WM, Kehn PJ et al. The vitronectin rceptor serves as an accessory molecule for the activation of a subset of $\gamma\delta$ T cells. J Exp Med 1991; 173:231-40.
146. Vuori K, Ruoslahti E. Association of insulin receptor substrate-1 with integrins. Science 1994; 266:1576-8.
147. Dedhar S. Novel functions for calreticulin: interaction with integrins and modulation of gene expression? Trends Biochem Sci 1994; 19:269-71.
148. Lawrence MB, Springer TA. Neutrophils roll on E-selectin. J Immunol 1993; 151:6638-46.
149. Bevilacqua MP. Endothelial-Leukocyte adhesion molecules. Annu Rev Immunol 1993; 11:767-804.
150. Fidler IJ. The biology of human cancer metastasis. Acta Oncol 1991; 30:668-75.
151. McCormik BA, Zetter BR. Adhesive interactions in angiogenesis and metastasis. Pharmac Ther 1992; 53:239-60.
152. Brodt P. Selection of a highly metastatic liver-colonizing subpopulation of Lewis lung carcinoma variant H-59 using murine hepatocyte monolayers. Clin Exp Metastasis 1989; 7:525-39.
153. Lafrenie RM, Gallo S, Podor TJ et al. The relative roles of vitronectin receptor, E-selectin and $\alpha_4\beta_1$ in cancer cell adhesion to interleukin-1-trated endothelial cells. Eur J Caner 1994; 30A:2151-8.
154. Picker LJ, Butcher EC. Physiological and molecular mechanisms of lymphocyte homing. Ann Rev Immunol 1992; 10:561-91.
155. Schnapp LM, Hatch N, Ramos DM et al. The human integrin $\alpha_8\beta_1$ functions as a receptor for tenascin, fibronectin and vitronectin. J Biol Chem 1995; 270:23196-23202.
156. Giancotti FG, Mainiero F. Integrin-mediated adhesion and signaling in tumorgenesis. Biochim Biophys Acta 1994; 1198:47-64.
157. Karecla PI, Bowden SJ, Green SJ et al. Recognition of E-cadherin on epithelial cells by the mucosal T cell integrin α(M290) β7 ($\alpha_E\beta_7$). Eur J Immunol 1995; 25:853-856.

CHAPTER 5

Cell Adhesion Molecules of the Immunoglobulin Superfamily

Chi-Hung Siu

INTRODUCTION

Aside from molecules related to the immune system, primary structures predicted from a number of cell adhesion molecule (CAM) cDNA sequences reveal regions containing significant sequence similarities with immunoglobulin (Ig) domains.[1-3] CAMs containing one or more Ig-like domains are considered members of the Ig superfamily of recognition molecules. These CAMs are often concentrated at regions of cell-cell contact and can also associate with specialized junctional complexes between cells (Fig. 5.1). CAMs in the Ig superfamily are capable of undergoing homophilic and heterophilic interactions with a variety of ligands[1] and function in a wide range of physiological processes, including embryonic and neural development,[3,4] inflammation and wound healing,[5-7] and oncogenesis.[8] Recently, several membrane-associated tyrosine kinases and protein tyrosine phosphatases have been found to contain Ig-like domains,[9-12] suggesting a direct link between cell-cell adhesion and certain signaling pathways.

Cell Adhesion and Invasion in Cancer Metastasis, edited by Pnina Brodt.

STRUCTURAL FEATURES OF CAMS IN THE Ig SUPERFAMILY

Ig CAMs contain one or more Ig-like domains extending from the surface of cells. Secondary and tertiary structure predictions suggest that these Ig-like segments adopt a structure characteristic of Ig and Ig-related domains, which are known to consist of β strands folded into two antiparallel β sheets tightly packed against each other (Fig. 5.2). The β strands generally contain alternating hydrophobic and hydrophilic amino acid residues, with the hydrophobic side-chains pointing inward and the hydrophilic ones pointing outward on the surface of the globular structure. The

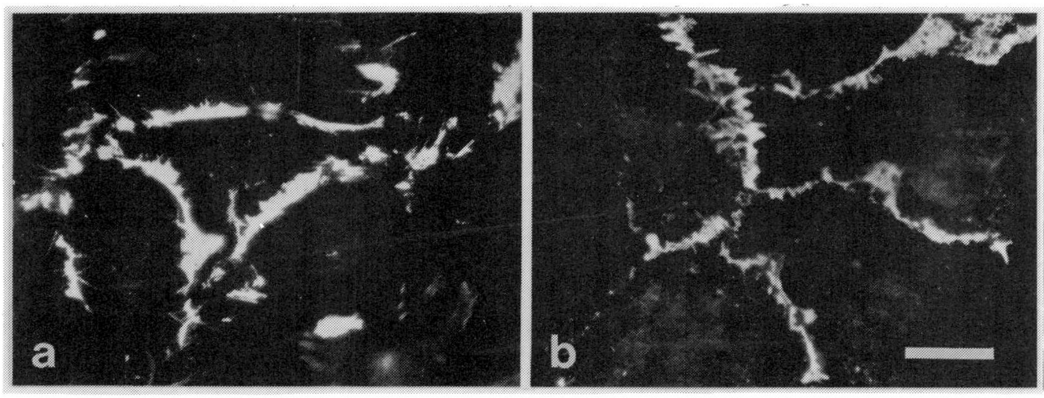

Fig. 5.1. Epifluorescence micrographs of CAM labeling of cell-cell contacts. (a) TM4 cells were stained indirectly using a rabbit anti-NCAM antiserum. (b) Human umbilical cord endothelial cells were stained with a rabbit anti-PECAM-1 antiserum. Note the strong CAM labeling at sites of cell-cell contact. Bar = 10 μm.

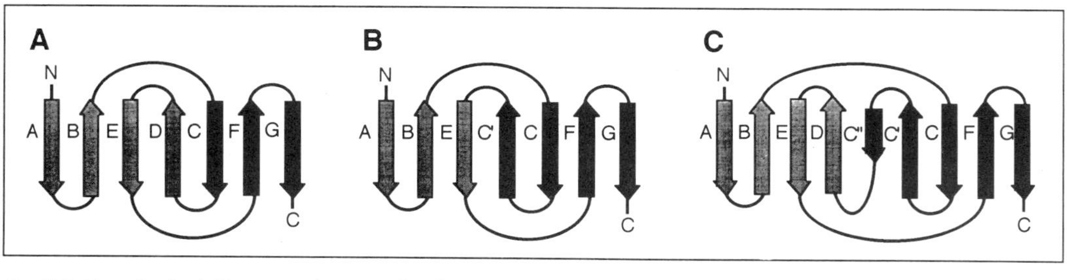

Fig. 5.2. Topological diagrams showing the three types of Ig fold: C1 (A), C2 (B), and V (C). The dotted arrows and the solid arrows represent, respectively, anti-parallel β strands hydrogen bonded to form either one of the two β sheets of the β barrel structure.

structure is stabilized through hydrogen bonding, hydrophobic interactions in the core, and a disulfide bond between the B and F strand of the two β sheets. Although the disulfide bridge was once considered a hallmark of Ig domains, several exceptions have been reported recently.[13,14]

Conserved structural characteristics and sequence homology are used to classify the Ig domains. Ig domains show two related folding patterns: the variable or V-type fold and the constant or C-type fold (Fig. 5.2). The V-domain generally consists of nine β strands with alternating hydrophobic residues. There are only seven β strands in the C-domain. The C-type fold is constituted by two β sheets comprised of β strands ABED and GFC, respectively. The V-type fold is distinguished from the C-type fold by the presence of an extra loop in the middle of the domain. This loop is made up of the C' and C'' β strands of the V-domain. Further analysis of sequence homology among Ig-like sequences has revealed the presence of a new class of Ig domains and the C-domains have been subdivided into the C1-SET and C2-SET.[1] Members of the C2-SET usually have a smaller disulfide loop, with the E and F strands showing greater resemblance with the corresponding β strands of V-domains. Moreover, recent crystallographic studies of C2 structures show that strand C', which corresponds sequentially to the D strand of the constant Ig domain, is flipped over to the β sheet consisting of strands GFC and becomes hydrogen-bonded with the C strand.[15,16] Most Ig domains in Ig CAMs belong to the C2-SET.[1] Although they all share some highly conserved sequences, no residue has been found to be invariant in all Ig-related domains.

Ig CAMs have a wide tissue distribution and are involved in a variety of cellular functions, most notably in neuronal differentiation and leukocyte-endothelial cell adhesion. Neural CAMs that belong to the Ig superfamily show diverse structural complexities (Fig. 5.3). The myelin Po glycoprotein is relatively simple containing a single V-type Ig domain,[1,17,18] whereas L1, Tag-1 and Nr-CAM, contain six Ig domains of the C2-SET.[19-23] The neural cell adhesion molecule NCAM and myelin associated glycoprotein (MAG) contain five Ig domains.[24,25] Ig CAMs that are expressed on endothelial cells include ICAM-1, ICAM-2, VCAM-1 (also known as INCAM-110), and PECAM-1. ICAM-1 is a 90 kDa membrane glycoprotein containing five Ig domains,[26] while

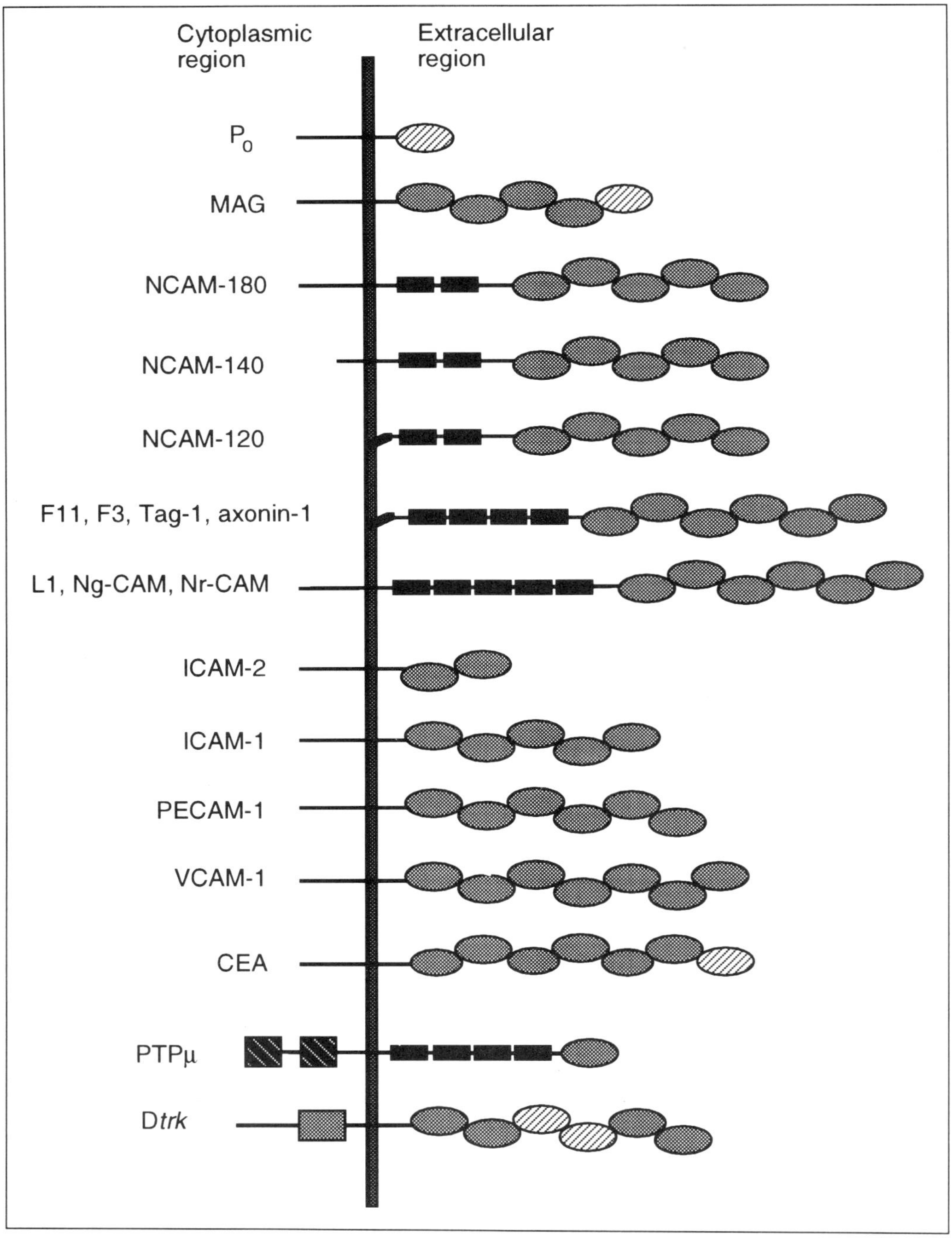

Fig. 5.3. Schematic drawings for CAMs in the Ig superfamily. The V-type Ig domains are shown in hatched ovals and the C2-type Ig domains are shown in stippled ovals. The solid rectangles represent fibronectin type III repeats, the thick slanted bar the phosphatidyl inositol anchor to the membrane, the hatched squares the phosphatase domains in PTPμ, and the stippled rectangle the tyrosine kinase domain in Dtrk.

ICAM-2 contains only two Ig domains.[27] Both ICAM-1 and ICAM-2 are involved in inflammatory reactions and bind to LFA-1, a member of the β_2 integrin family.[28] VCAM-1 contains seven Ig domains[29] and is known to bind to the α_4 subunit of VLA-4, the $\alpha_4\beta_1$ integrin.[6] The platelet-endothelial cell adhesion molecule PECAM-1 contains six Ig domains and is a major component of the endothelial cell-cell junction.[30,31]

Typically, sequences within a single Ig domain are encoded by a single exon. Occasionally, an Ig domain may consist of sequences encoded by two separate exons, as in the case of NCAM.[32,33] Internal repeats are apparent in many Ig CAMs, implying that they may arise by gene duplication. For instance, domains 1, 2 and 3 of VCAM-1 show a high degree of structural similarity with domains 4, 5 and 6, respectively, suggesting that the basic unit constituted by domains 1, 2 and 3 may have been duplicated to give rise to domains 4, 5 and 6.[29] Another example is the carcinoembryonic antigen (CEA), a well-known human tumor marker, which is made up of seven contiguous Ig domains.[8] The aminoterminal V-type domain is followed by six domains of the C2-SET. The latter domains exist as three homologous pairs, which are > 80% homologous at the nucleotide sequence level, implying duplication and divergence from a primordial unit.

In addition to Ig-like domains, CAMs in the Ig superfamily may contain other protein modules and motifs (Fig. 5.3). Further complexities arise from variations as a result of stage-specific and tissue-specific post-translational modifications and the presence of alternatively spliced isoforms. These aspects are best illustrated by NCAM, since its structure has been defined in considerable detail.[24] The extracellular domain of NCAM consists of five contiguous Ig-like domains of the C2-SET, each with a characteristic intrachain disulfide bond. The fifth Ig-like domain is followed by a proline-rich hinge structure and then a stalk region, which contains sequences homologous to the fibronectin type III repeats. A consensus heparin sulfate binding domain is present in the second Ig domain[34,35] and a sequence in Ig domain 4 shows homology with the C-type lectin motif.[36] Electron microscopy studies have revealed that NCAM has a rod-shaped structure with a bend in the middle,[37,38] confirming a linear arrangement of its protein modules and the presence of the hinge region. Alternative splicing of mRNA transcribed from a single gene gives rise to a large number of NCAM isoforms.[33] The three major neural NCAM forms are NCAM-180 (M_r ~180 kDa), NCAM-140 (M_r ~140 kDa), and NCAM-120 (M_r ~120 kDa), which have similar amino-terminal extracellular domains but truncated carboxyl-terminal domains. Whereas NCAM-180 and NCAM-140 are integral membrane proteins, NCAM-120 is attached to membranes via a glycosylphosphatidylinositol anchor.[39,40] NCAM is further diversified by post-translational modifications, including N-glycosylation, phosphorylation of serine and threonine residues in the cytoplasmic domain, fatty acid acylation, and sulfation of N-linked oligosaccharides.[24] The most striking modification is the α2,8-linked polysialic acid, the amount of which varies with the tissue and is regulated during development.[41,42] The combination of different structural motifs and post-translational modifications gives rise to a great variety of potential mechanisms to augment or modulate adhesive events between cells.

MECHANISMS OF CELL-CELL ADHESION MEDICATED BY Ig CAMS

Protein-Protein Interactions Involved in Ig CAM-Mediated Cell Adhesion

Cell-cell adhesion mediated by Ig CAMs can be a fairly complicated event. The presence of internal repeats in most Ig CAMs implies the presence of multiple contact sites along the length of the molecule. These sites may vary in their adhesive strength, thus providing a mechanism for fine-tuning the adhesive interaction. It is conceivable that certain low affinity sites may be utilized during cell migration and movement, while high affinity sites are involved in the formation of stable cellular structures. On the other hand, the repeated domains may have diverged to provide unique interacting sites for different ligands. The matter is further complicated by the simultaneous expression of several CAMs on the same cell. Lateral interactions between different CAMs in the same membrane plane may provide another level of regulation of the intercellular adhesiveness. As a first step to elucidate the mechanism of CAM function, it is necessary to define the binding sites in these molecules. To date, relatively few Ig CAMs have been analyzed in detail. The heterophilic binding site of ICAM-1 for LFA-1 has been localized to the first amino-terminal Ig domain,[26] while its binding site for Mac-1 maps to the third Ig domain.[43] On the other hand, VLA4 binds to both Ig domain 1 as well as domain 4 of VCAM-1.[44] The specific sequences involved remain to be determined.

NCAM is known to undergo homophilic binding with NCAM on apposing cells.[45-47] On the basis of monoclonal antibody mapping, the homophilic binding site of NCAM has been proposed to reside in a region corresponding to Ig domains 3 and 4.[24,48] Recently, a decapeptide sequence (243-KYSFNYDGSE-252) located in the third Ig domain of chick NCAM has been identified as a homophilic binding site.[49,50] The first six amino acids of this sequence assume a pattern of alternating hydrophilic and hydrophobic residues, which is typical of a β strand structure, while the carboxyl-terminal residues have a high probability of adopting a β turn structure. By analogy with the known crystal structures of Ig-related domains,[15,16,51] this region corresponds to the C' strand of the β barrel structure. Binding studies indicate that this sequence interacts isologously with the same sequence on an apposing NCAM.[52] The aromatic residues, as well as the charged residues, in the β strand region appear to play a vital role in homophilic interactions.[50]

While the sequence of this homophilic binding site is unique to NCAM, it appears to share some structural features with the homophilic binding site of gp80, a CAM known to mediate Ca^{2+}-independent cell-cell adhesion in the cellular slime mold *Dictyostelium discoideum.*[53,54] gp80 shows limited sequence and structural homology with NCAM and IgA, suggesting that it may also belong to the Ig superfamily of recognition molecules.[55-57] The gp80 homophilic binding site consists of the octapeptide sequence YKLNVNDS. Interestingly, this sequence is also predicted to adopt a β-strand conformation with a β turn at its carboxyl-terminus.[58] The two charged residues (Lys and Asp) are crucial in the initial electrostatic interactions between gp80 molecules. Binding between the two molecules is further stabilized by hydrophobic interactions and the formation of hydrogen bonds.[58]

Whereas both NCAM and gp80 mediate cell-cell adhesion by isologous binding, homophilic interactions of CEA are achieved by heterologous binding.[59] Two antiparallel CEA molecules interact by double reciprocal bonds between the Ig V-type amino-terminal domain of one molecule and the internal C-type domain

A3 of the apposing molecule. The exact sequence and structure of the two binding sites have not been analyzed. It would be of interest to determine if the C' strands of these two Ig domains participate in direct binding as in the case of NCAM[50] and CD2.[51]

Protein-Carbohydrate Interactions Involved in Ig CAM-Mediated Cell Adhesion

A consensus sequence for glycosaminoglycan binding has been detected in several Ig CAMs.[30,33] These sites can mediate the binding of a CAM-expressing cell to proteoglycans expressed on another cell. It may also facilitate the attachment of cells to the extracellular matrix. A heparan sulfate-binding site of NCAM has been localized to the second Ig domain,[48,60] and more recently to a 17-amino acid segment (131-IWKHKGRDVILKKDVRF-147) located within this domain.[33,34] It contains two stretches of basic amino acid residues and corresponds to the C strand and the turn structure linking the C and C' strands of the predicted Ig fold. The segment has been proposed to form an α-helical structure, with the positively charged residues lining up on the same side of the helix.[34,61] This positively charged surface then acts as an initial attachment site for the negatively charged heparin molecule. Since β barrel structures are known to be relatively stable, it will be of interest to determine whether the heparan sulfate binding site normally exists in helical form or whether the binding of heparan sulfate can induce conformational changes in this region of the molecule.

Recently, a B cell-restricted CAM, CD22β, which contains seven extracellular Ig domains,[62] has been found to behave like a lectin. CD22β binds to a large variety of glycoproteins bearing α2,6-linked sialic acids, which appear to be an absolute requirement in the binding reaction.[63,64] Protein-carbohydrate interactions have also been implicated in lateral interactions between NCAM and L1 and such interactions may have an effect on L1-L1 binding.[65] This *cis*-association is mediated apparently by interactions between NCAM Ig domain 4 and oligomannosidic carbohydrates of L1.[36,66] Sequence analysis reveals sequence homology between NCAM Ig domain 4 and carbohydrate recognition domains of animal C-type lectins. Both oligomannosidic glycans and peptides comprising part of the C-type lectin consensus sequence in domain 4 have been reported to inhibit NCAM-L1 association as well as neurite outgrowth from cerebellar neurons.[36]

SIGNALING PATHWAY INDUCED BY Ig CAMS

A substantial amount of evidence has accumulated supporting a role of Ig CAMs in initiating signal transduction cascades that lead to gene regulation and cell differentiation. The pathway involving several neural Ig CAMs has been studied extensively in recent years.[67] Fibroblasts expressing a variety of neural CAMs by gene transfer are capable of promoting neurite outgrowth from primary neurons in coculture systems.[21,68-72] It has become evident that this signaling cascade is initiated by cell-cell contact, since mutations in the homophilic binding site of NCAM abolishes its ability to promote neurite outgrowth.[73] Several other domains of NCAM have also been shown to influence neurite outgrowth and fasciculation.[74-76] It is evident that NCAM is a multi-domain molecule which can serve a variety of functions. Intermolecular interactions involving one domain may have subtle effects on the function of other domains. Exactly how signals are transmitted from the extracellular domain to the cytoplasmic side of the cell is unclear. It has been suggested that tyrosine phosphorylation-dephosphorylation is involved in an early step in the signaling pathway.[77,78] Recent studies using *fyn*$^-$ and *src*$^-$ mice showed that NCAM-dependent neurite outgrowth involves p59^{fyn} in its signaling pathway,[79] while pp60$^{c\text{-}src}$ is an essential component in the pathway initiated by L1.[80] However, many of the downstream events appear to be shared by the signaling pathways triggered by NCAM, L1 and Thy-1 to elicit the neurite outgrowth response.[72,81,82] Antagonists that block L- and N-type Ca^{2+} channels have been shown to inhibit the contact-dependent neurite-outgrowth response,[72] indicating a central role for intracellular Ca^{2+} in the second messenger pathway. Pharmacological studies also indicate that a pertussis toxin-sensitive heterotrimeric G protein acts upstream of the calcium channel activation step.[81] Downstream events probably involve multiple protein phosphorylation events.[72,77,83]

In recent years, many membrane components that contain Ig-like domains have been discovered. The recent cloning of the D*trk* gene,[12] a *Drosophila* homologue of the mammalian neurotropic receptors, is of particular interest (Fig. 5.3). This gene codes for a transmembrane protein with six Ig-like domains on the extracellular surface and a tyrosine kinase domain on the cytoplasmic side of the molecule. When expressed in transfected cells, the D*trk* protein mediates cell aggregation and induces tyrosine phosphorylation.

Moreover, it was recently reported that the protein tyrosine phosphatase PTPμ, which contains an Ig-like domain in its extracellular region, can mediate cell-cell adhesion when expressed in Sf9 cells.[10] Although it is not clear whether the D*trk* protein and PTPμ behave as CAMs in vivo, these observations do raise the possibility that signaling by Ig CAMs may be mediated by its association with adaptors that contain kinase or phosphatase activity.

CONCLUDING REMARKS

The important role of Ig CAMs in normal and pathological development has only been recognized in recent years. At present, the use of CAMs and their antibodies in treatment of diseases is being actively investigated in many laboratories. Future development of these and other related therapeutic reagents will depend on our knowledge of the structure/function relationships of these CAMs. More functional domains need to be mapped and further molecular details can be resolved by crystallographic and NMR studies. As we seek to understand more precisely the physiological role of different CAMs and their mechanisms of action, the interactions of CAMs with cytoskeletal elements and components of signaling pathways will constitute another major focus in CAM research.

ACKNOWLEDGMENTS

Research carried out in our laboratory was supported by the Medical Research Council of Canada.

REFERENCES

1. Williams AF, Barclay AN. The immunoglobulin superfamily: domains for cell surface recognition. Annu Rev Immunol 1988; 6: 381-405.
2. Hunkapiller T, Hood L. Diversity of the immunoglobulin gene superfamily. Adv Immunol 1989; 44:1-63.
3. Edelman GM, Crossin KA. Cell adhesion molecules: implications for a molecular histology. Ann Rev Biochem 1991; 60:155-90.
4. Sonderegger P, Rathjen FG. Regulation of axonal growth in the vertebrate nervous system by interactions between glycoproteins belonging to two subgroups of the immunoglobulin superfamily. J Cell Biol 1992; 119:1387-94.
5. Wawryk SO, Novotny JR, Wicks IP et al. The role of the LFA-1/ICAM-1 interaction in human leukocyte homing and adhesion. Immunol Rev 1989; 108:135-61.

6. Elices MJ, Osborn L, Takada Y et al. VCAM-1 on activated endothelium interacts with the leukocyte integrin VLA-4 at a site distinct from the VLA4/fibronectin binding site. Cell 1990; 60:577-84.
7. Carlos TM, Harlan JM. Leukocyte-endothelial adhesion molecules. Blood 1994; 84:2068-101.
8. Benchimol S, Fuks A, Jothy S et al. Carcinoembryonic antigen, a human tumor marker, functions as an intercellular adhesion molecule. Cell 1989; 57:327-34.
9. Streuli M, Krueger N, Tsai A et al. A family of receptor-linked protein tyrosine phosphatases in humans and *Drosophila*. Proc Natl Acad Sci USA 1989; 86:8698-702.
10. Brady-Kalnay SM, Flint AJ, Tonks NK. Homophilic binding of PTPμ, a receptor-type protein tyrosine phosphatase, can mediate cell-cell aggregation. J Cell Biol 1993; 122:961-72.
11. Yarden Y, Escobedo JA, Kuang WJ et al. Structure of the receptor for platelet-derived growth factor helps define a family of closely related growth factor receptors. Nature 1986; 323:226-32.
12. Pulido D, Campuzano S, Koda T et al. D*trk*, a *Drosophila* gene related to the *trk* family of neurotrophin receptors, encodes a novel class of neural cell adhesion molecule. EMBO J 1992; 11:391-404.
13. Killeen N, Moessner R, Arvieux J et al. The MRC OX-45 antigen of rat leukocytes and endothelium is in a subset of the immunoglobulin superfamily with CD2, LFA-3 and carcinoembryonic antigens. EMBO J 1988; 7:3087-91.
14. Maddon PJ, Littman DR, Godfrey DE et al. The isolation and nucleotide sequence of a cDNA encoding the lymphocyte protein T4: a new member of the immunoglobulin family. Cell 1985; 41:93-104.
15. Wang J, Yan Y, Garrett TPJ et al. Atomic structure of a fragment of human CD4 containing two immunoglobulin-like domains. Nature 1990; 348:411-8.
16. Ryu S, Kwong PD, Truneh A et al. Crystal structure of an HIV-binding recombinant fragment of human CD4. Nature 1990; 348:419-26.
17. Lemke G, Axel R. Isolation and sequence of a cDNA encoding the major structural protein of peripheral myelin. Cell 1987; 40:501-8.
18. Filbin MT, Walsh FS, Trapp JA et al. Role of myelin Po-protein as a homophilic adhesion molecule. Nature 1990; 344:871-2.
19. Hlavin ML, Lemmon V. Molecular structure and functional testing of human L1 CAM: an interspecies comparison. Genomics 1991; 11:416-23.
20. Grumet M. Structure, expression and function of Ng-CAM, a member of the immunoglobulin superfamily involved in neuron-neuron and neuron-glia adhesion. J Neurosci Res 1992; 31:1-13.

21. Furley AJ, Morton SB, Manalo D et al. The axonal glycoprotein TAG-1 is an immunoglobulin superfamily member with neurite outgrowth-promoting activity. Cell 1990; 61:157-70.
22. Zuellig RA, Rader C, Shroeder A et al. The axonally secreted cell adhesion molecule, axonin-1. Eur J Biochem 1992; 204:453-63.
23. Mauro VP, Krushel LA, Cunningham BA et al. Homophilic and heterophilic binding activities of Nr-CAM, a nervous system cell adhesion molecule. J Cell Biol 1992; 119:191-202.
24. Cunningham BA, Hemperly JJ, Murray BA et al. Neural cell adhesion molecule: Structure, immunoglobulin-like domains, cell surface modulation, and alternative RNA splicing. Science 1987; 236:799-806.
25. Lai C, Brow MA, Nave KA et al. Two forms of 1B236/myelin-associated glycoprotein, a cell adhesion molecule for postnatal neural development, are produced by alternative splicing. Proc Natl Acad Sci USA 1987; 84:4337-41.
26. Staunton DE, Martin SD, Stratowa C et al. Primary structure of ICAM-1 demonstrates interaction between members of the immunoglobulin and integrin supergene family. Cell 1988; 52:925-33.
27. Staunton DE, Dustin ML, Springer TA. Functional cloning of ICAM-2, a cell adhesion ligand for LFA-1 homologous to ICAM-1. Nature 1989; 339:61-4.
28. Springer TA. Adhesion receptors of the immune system. Nature 1990; 346:425-33.
29. Hession C, Tizard R, Vassallo C et al. Cloning of an alternate form of vascular cell adhesion molecule-1 (VCAM1). J Biol Chem 1991; 266:6682-5.
30. Newman PJ, Berndt MC, Gorski J et al. PECAM-1 (CD31): cloning and relation to adhesion molecules of the immunoglobulin gene superfamily. Science 1990; 247:1219-22.
31. Albelda SM, Muller WA, Buck CA et al. Molecular and cellular properties of PECAM-1 (endoCAM/CD31): a novel vascular cell-cell adhesion molecule. J Cell Biol 1991; 114:1059-68.
32. Barbas JA, Chaix JC, Steinmetz M et al. Differential splicing and alternative polyadenylation generates distinct NCAM transcripts and proteins in the mouse. EMBO J 1988; 7:625-32.
33. Walsh FS, Doherty P. NCAM gene structure and function. Semin Neurosci 1991; 3:271-84.
34. Cole GJ, Akeson R. Identification of a heparin binding domain of the neural cell adhesion molecule NCAM using synthetic peptides. Neuron 1989; 2:1157-65.
35. Reyes AA, Akeson R, Brezina L et al. Structural requirements for neural cell adhesion molecule-heparin interaction. Cell Regulation 1990; 1:567-76.

36. Horstkorte R, Schachner M, Magyar JP et al. The fourth immunoglobulin-like domain of NCAM contains a carbohydrate recognition domain for oligomannosidic glycans implicated in association with L1 in neurite outgrowth. J Cell Biol 1993; 121:1409-21.
37. Hall AK, Rutishauser U. Visualization of neural cell adhesion molecule by electron microscopy. J Cell Biol 1987; 104:1579-86.
38. Becker JW, Erickson HP, Hoffman S et al. Topology of cell adhesion molecules. Proc Natl Acad Sci USA 1989; 86:1088-92.
39. He HT, Barbet J, Chaix JC et al. Phosphatidylinositol is involved in the membrane attachment of N-CAM 120, the smallest component of the neural cell adhesion molecule. EMBO J 1986; 5:2489-94.
40. Hemperly JJ, Edelman GM, Cunningham BA. cDNA clones of the neural cell adhesion molecule (NCAM) lacking a membrane-spanning region consistent with evidence for membrane attachment via a phosphatidylinositol intermediate. Proc Natl Acad Sci USA 1986; 83:9822-26.
41. Edelman GM. Cell adhesion molecules. Science 1983; 219:450-7.
42. Rutishauser U, Acheson A, Hall AK et al. The neural cell adhesion molecule (NCAM) as a regulator of cell-cell interactions. Science 1988; 240:53-57.
43. Diamon MS, Staunton DE, Marlin SD et al. Binding of the integrin Mac-1 (CD11b/CD18) to the third immunoglobulin-like domain of ICAM-1 (CD54) and its regulation by glycosylation. Cell 1991; 65:961-71.
44. Osborn I, Vassallo C, Benjamin CD. Activated endothelium binds lymphocytes through a novel binding site in the alternately spliced domain of VCAM-1. J Exp Med 1991; 176:99-108.
45. Hoffman S, Edelman GM. Kinetics of homphilic binding by embryonic and adult forms of the neural cell adhesion molecule. Proc Natl Acad Sci USA 1983; 80:5762-6.
46. Edelman GM, Murray BA, Mege R et al. Cellular expression of liver and neural cell adhesion molecules after transfection with their cDNAs results in specific cell-cell binding. Proc Natl Acad Sci USA 1987; 84:8502-6.
47. Pizzey JA, Rowett LH, Barton CH et al. Intercellular adhesion mediated by human muscle neural cell adhesion molecule: effects of alternative exon use. J Cell Biol 1989;109:3465-76.
48. Frelinger III AL, Rutishauser U. Topography of NCAM structural and functional determinants. II. Placement of monoclonal antibody epitopes. J Cell Biol 1986; 103:1729-37.
49. Rao Y, Wu XF, Gariepy J et al. Identification of a peptide sequence involved in homophilic binding in the neural cell adhesion molecule NCAM. J Cell Biol 1992; 118:937-49.

50. Rao Y, Wu XF, Yip P et al. Structural characterization of a homophilic binding site in the neural cell adhesion molecule NCAM. J Biol Chem 1993; 268:20630-38.
51. Jones EY, Davis SJ, Williams AF et al. Crystal structure at 2.8 A resolution of a soluble form of the cell adhesion molecule CD2. Nature 1992; 360:232-39.
52. Rao Y, Zhao X, Siu CH. Mechanism of homophilic binding mediated by the neural cell adhesion molecule NCAM: evidence for isologous interaction. J Biol Chem 1994; 269:27540-8.
53. Muller K, Gerisch G. A specific glycoprotein as target of adhesion blocking Fab in aggregating *Dictyostelium discoideum*. Nature 1978; 274:445-9.
54. Siu CH, Lam TY, Choi HAC. Inhibition of cell-cell binding at the aggregation stage of *Dictyostelium discoideum* development by monoclonal antibodies directed against an 80,000-Dalton surface glycoprotein. J Biol Chem 1985; 16030-6.
55. Siu CH, Cho A, Choi AHC. The contact site A glycoprotein mediates cell-cell adhesion by homophilic binding in *Dictyostelium discoideum*. J Cell Biol 1987; 105:2523-33.
56. Matsunaga T, Mori N. The origin of the immune system: the possibility that immunoglobulin and cell adhesion molecules of chicken and slime mould are all related. Scand J Immunol 1987; 25:485-95.
57. Siu CH, Kamboj RK. Cell-cell adhesion and morphogenesis in *Dictyostelium discoideum*. Dev Genet 1990; 11:377-87.
58. Kamboj RK, Gariepy J, Siu CH. Identification of an octapeptide involved in homophilic interaction of the cell adhesion molecule gp80 of *Dictyostelium discoideum*. Cell 1989; 59:615-25.
59. Zhou H, Fuks A, Alcaraz G et al. Homophilic adhesion between Ig superfamily carcinoembryonic antigen molecules involves double reciprocal bonds. J Cell Biol 1993; 122:951-60.
60. Cole GJ, Loewy A, Cross NV et al. Topological localization of the heparin-binding domain of the neural cell adhesion molecule NCAM. J Cell Biol 1986; 103:1739-44.
61. Cardin AD, Weintraub HJR. Molecular modeling of protein of protein-glycosaminoglycan interactions. Arteriosclerosis 1988; 9:21-32.
62. Wilson GL, Fox CH, Fauci AS et al. cDNA cloning of the B cell membrane protein CD22: a mediator of B-B cell interactions. J Exp Med 1991; 173:137-46.
63. Stamenkovic I, Sgroi D, Aruffo A et al. The B lymphocyte adhesion molecule CD22 interacts with leukocyte common antigen CD45RO on T cells and α2,6 sialyltransferase, CD75, on B cells. Cell 1991; 66:1133-44.
64. Powell LD, Varki A. The oligosaccharide binding specificities of CD22β, a sialic acid-specific lectin of B cells. J Biol Chem 1994; 269:10628-36.

65. Kadmon GA, Kowitz A, Altevogt P et al. The neural cell adhesion molecule N-CAM enhances L1-dependent cell-cell interactions. J Cell Biol 1990; 110:193-208.
66. Kadmon GA, Kowitz A, Altevogt P et al. Functional cooperation between the neural adhesion molecules L1 and NCAM is carbohydrate dependent. J Cell Biol 1990; 110:209-18.
67. Walsh FS, Doherty P. Factors regulating the expression and function of calcium-independent cell adhesion molecules. Curr Opin Cell Biol 1993; 5:791-96.
68. Gennarini G, Durbec P, Boned A et al. Transfected F3/F11 neuronal cell surface protein mediates intercellular adhesion and promotes neurite outgrowth. Neuron 1991; 6:595-606.
69. Doherty P, Cohen J, Walsh FS. Neurite outgrowth in response to transfected N-CAM changes during development and is modulated by polysialic acid. Neuron 1990; 5:209-19.
70. Kuhn TB, Stoeckli ET, Condrau MA et al. Neurite outgrowth on immobilized axonin-1 is mediated by a heterophilie interaction with LI(GH). J Cell Biol 1991; 115:1113-26.
71. Yazaki T, Miura M, Asou H et al. Myelin P_0 protein expressed in C6 cells promotes neurite outgrowth. Biomed Res 1991; 12:223-30.
72. Doherty P, Ashton SV, Moore SE et al. Morphoregulatory activities of NCAM and N-cadherin can be accounted for by G-protein dependent activation of L- and N-type neuronal calcium channels. Cell 1991; 67:21-33.
73. Sandig M, Rao Y, Siu CH. The homophilic binding site of the neural cell adhesion molecule NCAM is dirctly involved in promoting neurite outgrowth from cultured neural retinal cells. J Biol Chem 1994; 269:14841-8.
74. Pollerberg GE, Beck-Sickinger A. A functional role of the middle extracellular region of the neural cell adhesion molecule (NCAM) in axonal fasciculation and orientation. Dev Biol 1993; 156:324-40.
75. Doherty P, Moolenaar CECK, Ashton SV et al. The VASE exon downregulates the neurite growth-promoting activity of NCAM 140. Nature 1992; 356:791-93.
76. Frei T, von Bohlen und Halbach F, Wille W et al. Different extracellular domains of the neural cell adhesion molecule (N-CAM) are involved in different functions. J Cell Biol 1992; 118:177-94.
77. Atashi JR, Stephan SG, Klinz G et al. Neural cell adhesion molecules modulate tyrosine phosphorylation of tubulin in nerve growth cone membranes. Neuron 1992; 8:831-42.
78. Williams EJ, Walsh FS, Doherty P. Tyrosine kinase inhibitors can differentially inhibit integrin-dependent and CAM-stimulated neurite outgrowth. J Cell Biol 1994; 124:1029-37.
79. Beggs HE, Soriano P, Maness PF. NCAM-dependent outgrowth is inhibited in neurons from Fyn-minus mice. J Cell Biol 1994; 127:825-33.

80. Ignelzi MA, Miller DR, Soriano P et al. Impaired neurite outgrowth of src-minus cerebellar neurons on the cell adhesion molecule L1. Neuron 1994; 12:873-84.
81. Williams EJ, Doherty P, Turner G et al. Calcium influx into neurons can solely account for cell contact-dependent neurite outgrowth by transfected L1. J Cell Biol 1992; 119:883-92.
82. Doherty P, Singh A, Rimon G et al. Thy-1 antibody-triggered neurite outgrowth requires an influx of calcium into neurons via N-and L-type calcium channels. J Cell Biol 1993; 122:181-9.
83. Schuch U, Lohse MJ, Schachner M. Neural cell adhesion molecules influence second messenger systems. Neuron 1989; 3:13-20.

CHAPTER 6

SELECTINS

Tanya N. Mayadas

The initial step in leukocyte emigration during inflammation is the adherence of circulating leukocytes to the endothelial lining. This interaction is supported by a number of different constitutively expressed and inducible adhesion molecules on the surfaces of leukocytes and endothelial cells. Arrest and firm adhesion of leukocytes occurs through interaction between β_2 integrins, which are upregulated on activated leukocytes, and the endothelial inducible ligand ICAM-1. An additional pair of adhesion molecules, VCAM-1 on endothelial cells and VLA4 integrin present on monocytes and lymphocytes are also responsible for leukocyte-endothelial cell interaction.[1,2] However the specificity for leukocyte recruitment and subsequent local activation of leukocytes relies on transient interaction of inactivated leukocytes with the activated endothelium. This initial interaction of flowing leukocytes with the venular wall translates to leukocyte rolling and is mediated by the selectin family of adhesion molecules (Fig. 6.1). The three members of the selectin family, P-, E- and L-selectin were discovered respectively on platelets (and later on endothelium), endothelium and leukocytes and were named to emphasize the presence of the lectin domain as well as the selective expression and function of these molecules.

SELECTIN EXPRESSION

Expression of P- and E-selectin has been documented at sites of active inflammation in both animal and human tissue and both are receptors for neutrophils, monocytes, eosinophils and a subset

Cell Adhesion and Invasion in Cancer Metastasis, edited by Pnina Brodt.

of lymphocytes.[3] Despite the structural and functional similarity, the expression of these two selectins on the endothelium is different. P-selectin, originally described on the surface of activated platelets, was subsequently shown also to be stored in the Weibel-Palade bodies of endothelial cells and the α-granules of platelets. From these intracellular stores, P-selectin can be redistributed within minutes to the cell surface upon stimulation with various inflammatory mediators and thrombogenic substances which lead to Weibel-Palade body degranulation. These secretagogues include histamine, thrombin, C5b9, hydrogen peroxide, phorbol myristate

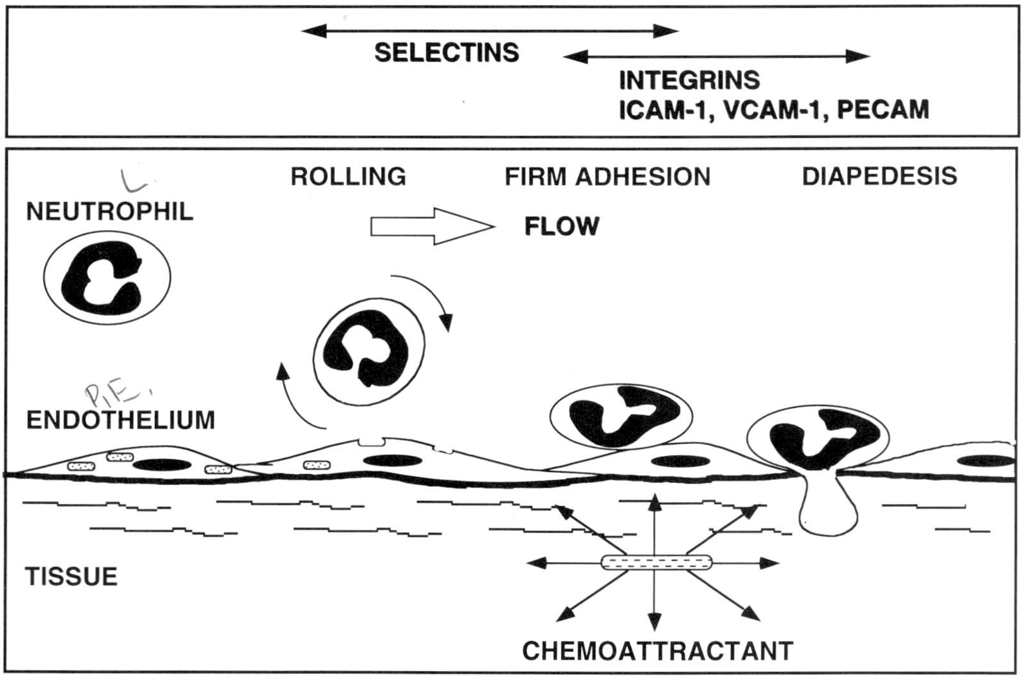

Fig. 6.1. Multistep model of neutrophil recruitment to an inflammatory site: Circulating neutrophils roll on the vessel wall by engaging selectins, P- and E-selectin on the endothelium and L-selectin on the neutrophil. Preformed P-selectin is stored in endothelial rod-shaped storage granules, the Weibel-Palade bodies, and is rapidly translocated to the cell surface. E-selectin as well as P-selectin are cytokine inducible and are present on the endothelial surface approximately 4 hrs after cytokine stimulation. L-selectin is constitutively expressed on the surface of neutrophils and is shed upon neutrophil activation. Neutrophil rolling is a prerequisite to firm adhesion and diapedesis which is supported by the integrins on the activated neutrophils and members of the IgG superfamily on the endothelium. This multistep model also applies to other leukocyte subpopulations. The specificity for leukocyte subpopulations is achieved by elaboration of different combinations of leukocyte adhesion molecules and their counter-receptors, and the presence of chemokines that have specificity for leukocyte subsets.

acetate (PMA), and the Ca ionophore A23187.[4,5] Depending on the stimulus, cell surface expression may be short-lived and may constitute a pathway to regulate P-selectin expression at sites of vascular injury. Histamine treatment of endothelial cells leads to re-endocytosis of surface P-selectin within 15 minutes[6] and recycling of the molecule into Weibel-Palade bodies.[7] On the other hand, hydrogen peroxide treatment leads to prolonged P-selectin surface expression perhaps through impairment of the endocytosis machinery.[8] Cytokines have recently been reported to regulate P-selectin expression on endothelial cells. P-selectin message levels increased in mice after treatment with LPS,[9] and de novo P-selectin synthesis was upregulated upon TNFα or IL-1 activation of endothelioma cell lines.[10] Maximal surface expression occurred 4 hrs after TNFα stimulation.[11] This coincides with the time course reported for E-selectin surface expression. E-selectin's synthesis is induced de novo upon cytokine activation of endothelial cells with IL-1β and TNFα. It is present on the surface 4-6 hrs after endothelial cell activation and declines to basal levels by 24-48 hrs.[12] Like P-selectin, surface E-selectin is internalized after 20 minutes but it is redirected into lysosomes.[13]

L-selectin is constitutively expressed on lymphocytes, monocytes and neutrophils and is shed upon leukocyte activation. L-selectin is concentrated on the microvilli projections of unstimulated neutrophils which is advantageous for early contact of L-selectin with its endothelial counter-receptor.[14]

SELECTIN STRUCTURE

The genes for all three selectins are clustered within a 300 kb segment on both human and murine chromosome 1 and are closely linked to coagulation factor V and to genes for complement binding proteins. The selectins are integral membrane proteins composed of an N-terminal C-type lectin domain, EGF domain, variable number of complement repeats, a transmembrane domain and a cytoplasmic tail.[3]

Both the lectin and EGF domains are involved in ligand binding. This has been established in adhesion assays using recombinant fusion proteins lacking one or more domains or selectin chimeras carrying domains and by mapping of epitopes recognized by functional blocking antibodies.[15] The role of the complement domain is less well defined although deletion of this domain in

L-selectin leads to diminished adhesive function due to a concomitant decrease in the functionality of the lectin domain.[16] It is not known whether the complement-like repeats bind complement proteins. The cytoplasmic domain has been shown to mediate intracellular trafficking and modulate adhesive functions. The P-selectin cytoplasmic tail is highly conserved between species and contains determinants which target P-selectin into storage granules of endothelial cells.[17] This domain does not appear to be required for fusion of the storage granules with the plasma membrane since platelets from P-selectin deficient mice are still capable of releasing their alpha granules upon thrombin stimulation.[18] The cytoplasmic domain is however required for the internalization of surface P-selectin.[7] Phosphorylation of the cytoplasmic tail occurs upon platelet[19,20] or endothelial cell[20] activation with thrombin or PMA respectively. The functional relevance of this modification is not known since soluble P-selectin lacking the cytoplasmic domain still binds leukocytes.[21] Phosphorylation however may indirectly modulate P-selectin adhesive function either by promoting conformational changes that affect ligand binding or clustering of P-selectin at the surface. The cytoplasmic tail of L-selectin is important in supporting L-selectin mediated leukocyte rolling and lymphocyte adhesion to high endothelial venules (HEV) of peripheral lymph nodes although it is not required for carbohydrate recognition since the activity of the lectin domain remained unaffected when the cytoplasmic domain was deleted. The cytoplasmic tail may modulate receptor avidity by controlling cytoskeletal interactions since an intact microfilament system was found to be required for L-selectin mediated adhesion.[22,23]

CARBOHYDRATE LIGANDS

The Ca^{2+}-dependent lectin domains of all three selectins recognize Sialyl-Le^x, a sialylated, fucosylated tetrasaccharide, as well as a number of related carbohydrates. Sialyl-Le^x is highly expressed on neutrophils and monocytes[24,25,26] and on a small percentage of blood lymphocytes.[27] Interestingly, certain cancer cells contain both Sialyl-Le^x and Sialyl-Le^a suggesting a role for selectins in cancer metastasis.[28,29,30] Unrelated to these sialylated ligands, P-selectin has also been shown to recognize sulfatides, specifically on tumor cell lines, through a Ca^{2+} independent mechanism.[31] As is characteristic of many lectins, the selectins bind a variety of carbohydrate ligands in vitro. The relative binding avidity of the selectins for Sialyl-Le^x and its isomer Sialyl-Le^a differ and may be a contributing factor to the specificity of the individual selectins for their ligands.[32] In this section only possible biologically relevant ligands will be discussed although several other diverse oligosaccharides have been shown to bind selectins. For a recent discussion of this complex subject the reader is referred to Varki.[32]

Two sulfated L-selectin ligands were identified,[33] although several others may also exist. One of these ligands on murine peripheral lymph nodes high endothelial venules, designated Glycam 1, was recently cloned and identified as a novel mucin-like molecule which is potentially soluble.[34] The other potential L-selectin ligand, a mucin-like molecule designated CD34, stains the cell surface of peripheral lymph node endothelium.[35] Glycam-1 and CD34 require sialylation and sulfation of their O-linked side chains in order to bind L-selectin.[34,35] These two candidates are not thought to be the only ligands for L-selectin and neither of them (Glycam-1 or CD34) appear to support L-selectin-dependent leukocyte adhesion to activated endothelial cells.[35] An inducible L-selectin ligand on endothelium has been shown to exist.[36] Identification of such a ligand is currently an active area of investigation.

A functional ligand for P-selectin was cloned from an HL60 (myelomonocytic cell line) cDNA library.[37] P-selectin glycoprotein ligand-1 (PSGL), is a novel mucin-like transmembrane protein expressed in several monocyte cells lines and in freshly isolated PMN and monocytes. Although the soluble form of PSGL-1 bound to both P- and E-selectin, the protein component was determined to being important in defining PSGL-1 as a P-selectin ligand.[37] Another line of investigation revealed that L-selectin which is sialylated on neutrophils but not on lymphocytes may directly serve as a ligand for P- and E-selectin.[14] Abbassi et al[38] provided evidence for direct L- and E-selectin interactions under conditions of flow. The latter two observations however cannot account for the adhesion of PMA-stimulated neutrophils and HL60 cells which do not express L-selectin and yet adhere to P- and E-selectin.[3] E-selectin appears to share ligands with P-selectin but also recognize ligands that are specific only for E-selectin.[39] A 150 kDa myeloid glycoprotein purified from murine bone marrow cells[40] and a 250 kDa ligand on gamma/delta lymphocytes[41] were identified as specific E-selectin ligands. The 150 kDa protein was recently cloned (ESL-1). The murine CDNA was 94% homologous to chicken growth factor receptor, and the protein required fucosylation to support E-selectin sdhesion (Steegmaler et al[80]).

SELECTIN FUNCTION

The importance of neutrophil/endothelial cell adhesion in the development of the inflammatory response is underscored by two syndromes, Leukocyte Adhesion Deficiency I (LADI) and LADII. Patients with LADI, have a genetic defect in the common beta-2 integrin on leukocytes and suffer recurrent soft tissue infections, impaired pus formation and wound healing, gingivitis/periodontal disease, and demineralization of the alveolar bone.[42] Individuals with LADII, have a general deficiency in fucosyltransferases which leads to the absence of Sialyl-Le^x, an essential component of selectin ligands. They display multiple phenotypes including severe mental retardation, microencephaly, facial dysmorphism, dwarfism, recurrent infections and neutrophilia.[43,44] Neutrophils and monocytes from LADII patients show decreased leukocyte rolling[45] which strongly supports the role of these carbohydrate ligands in leukocyte emigration in vivo.

All three selectins are responsible for the initial leukocyte-endothelial cell interaction characterized by low affinity binding which allows inactivated leukocytes to resist the shear force exerted by blood flow and roll along the vessel wall.[2] Leukocyte rolling has been most often examined on endothelial monolayers in vitro under physiological wall shear rates or in vivo by intravital microscopy of the exteriorized mesentery of anesthetized mice or rats. An emerging concept is that the three selectins may have overlapping functions in leukocyte rolling in some inflammatory situations and may have sequential contributions in other situations. To better understand the relative roles of selectins in vivo in various aspects of leukocyte-endothelial cell interactions we developed P-selectin-deficient mice[18] using the method of gene targeting.[46,47] The most striking difference between the P-selectin-deficient and wild type mice was the complete absence of leukocyte rolling in the mesenteric venules immediately following exteriorization of this tissue. Our data initially appeared to contradict earlier in vitro[36,48-50] and in vivo[51,52] studies which implicated L-selectin as the primary mediator of leukocyte rolling. This difference was recently reconciled with the generation of L-selectin-deficient mice[53] in which leukocyte rolling was studied. L-selectin deficient mice had wild type levels of leukocytes rolling in the mesenteric preparations during the first 30 minutes of the experimental procedure. However, 70-120 minutes later the leukocyte flux in L-selectin deficient mice declined to 50% of wild type levels.[53] In combination with our studies on P-selectin-deficient mice,[18] these data suggest that the earliest phases of the acute response maybe P-selectin dependent whereas the later phases may be predominantly L-selectin dependent. To experimentally address this issue Ley et al[54] studied leukocyte rolling in L-selectin-deficient, P-selectin-deficient and wild type animals injected intravenously with functional blocking L- E- and P-selectin antibodies. Leukocyte rolling was studied by intravital microscopy of venules in the cremaster muscle of the animals shortly after exteriorization of the tissue. This study concluded that there is a sequential contribution of P- and L-selectin shortly following tissue trauma induced by exteriorization and a synergistic contribution of these two selectins at later time points. No role for E-selectin in leukocyte rolling was observed in this study. The role of P-selectin in the early phases of leukocyte rolling may be tissue-specific since there is considerable variation of constitutive P-selectin expression levels between tissues.[55,56] Leukocyte rolling was examined in the skin of P-selectin-deficient mice using a dorsal skin chamber preparation in the absence of externally applied inflammatory stimuli. In P-selectin-deficient mice leukocyte rolling was observed in the skin microvasculature although it was significantly reduced (approximately 3-fold) compared to wild type animals (Yamada et al[78]). This suggests that P-selectin is a major but not the only contributor to the rolling in the skin.

Under more protracted inflammatory conditions, the relative contributions of the selectins in rolling is more complex. For example, leukocyte rolling was unaffected in TNFα stimulated blood vessels of the cremaster muscle of P-selectin deficient mice[54] but could be completely blocked by a functional blocking monoclonal antibody to L-selectin. On the other hand, minimal leukocyte rolling was observed in the mesenteric venules of P-selectin deficient mice several hours after treatment of these mice with an inflammatory irritant, thioglycollate. Similar treatment of wild type animals led to extensive leukocyte rolling and adhesion (Johnson et al[79]). E-selectin appears to play a role in leukocyte rolling in IL-1 stimulated rabbit mesenteric venules.[57] In conclusion, the relative contributions of the selectins are likely to depend on the combination of stimuli present under different circumstances.

Historically, leukocyte rolling is seen as a prerequisite to leukocyte extravasation.[52] Leukocyte rolling may enhance leukocyte activation through inter- or intracellular signaling. In this regard P-selectin facilitates the activation of neutrophils by platelet activating factor (PAF), a membrane bound lipid mediator whose surface expression on endothelial cells coincides with P-selectin surface expression.[58] PAF may enhance neutrophil degranulation by priming leukocytes rolling on P-selectin.[59] Engagement of L-selectin leads to intracellular signaling that augments the oxidative burst of neutrophils.[60] Similar to the complex interplay of selectins in leukocyte rolling, leukocyte extravasation appears to be mediated by individual selectins in some cases and in other cases appear to have redundant or complementary functions. Antibodies to L-selectin or recombinant L-selectin-IgG chimeras diminished neutrophil recruitment to thioglycollate induced inflammation in the peritoneal cavity, to inflamed skin and lung.[61] In our studies with P-selectin-deficient mice, using the same thioglycollate induced peritonitis model, we found that neutrophil influx was severely compromised up to 2 hrs after thioglycollate injection with significant recovery of neutrophil influx by 4 hrs.[18] A similar, almost superimposible profile of neutrophil influx was obtained in thioglycollate treated L-selectin-deficient mice.[53] This suggests that both molecules function in parallel or that L- and P-selectin interact with each other. Simultaneous inhibition of both L- and P-selectin with functional blocking monoclonal antibodies completely blocked neutrophils influx in the thioglycollate model.[62] E-selectin-deficient mice on their own have both normal neutrophil influx into thioglycollate inflamed peritoneum and normal delayed type hypersensitivity responses. However injection of P-selectin antibody into E-selectin-deficient animals leads to a reduced neutrophil influx in the thioglycollate model and reduced contact hypersensitivity responsiveness compared to wild type animals injected with P-selectin antibody.[63] This suggests that E-selectin may have a redundant function with P-selectin in these inflammatory situations.

Another phenotype of P-selectin deficient mice that may be traced to an underlying defect in the interaction of neutrophils and possibly other leukocytes with the vascular endothelium is the neutrophilia observed in their circulating blood under basal conditions. P-selectin-deficient mice had 2.4-fold more circulating neutrophils than wild type mice.[18] The elevation of neutrophil counts is reminiscent of findings in LADI and LADII patients. In both conditions, leukocyte levels are elevated 5- to 20-fold.[42] There could be several reasons for this increase including a lack of leukocyte margination, an increase in mature neutrophil exit from the bone marrow or increased neutrophil clearance. Although the numbers of bone marrow progenitor cells of the myeloid and erythroid lineages were comparable in wild type and P-selectin null animals and leukocyte margination appeared normal, the neutrophil clearance in the P-selectin-deficient animals appeared to be impaired since the half life of injected labeled neutrophils was higher in P-selectin-deficient than in wild type animals (Johnson et al[79]). This suggests a role for P-selectin in the normal mechanism of neutrophil clearance which may occur upon neutrophil extravastion into tissues, or in spleen.

Although the role of P-selectin on the endothelium has been studied extensively, its role on platelets remains unclear. It is the only selectin found in platelets and activated platelets express 13-fold more P-selectin on their surface than activated endothelium.[56] P-selectin mediates interaction between platelets and leukocytes.[64,65] In vivo, P-selectin mediated platelet-leukocyte interaction in a developing thrombi promotes leukocyte accumulation and fibrin deposition.[66] P-selectin does not appear to be required for the production and survival, or for routine hemostasis since in P-selectin-depleted mice the numbers of platelets is normal and simple tests have not revealed any obvious defects in platelet release.[18] However, the role of P-selectin in platelet aggregation and other aspects of platelet function such as platelet clearance from the circulation remain to be analyzed.

In addition to L-selectin's role in neutrophil recruitment to activated endothelial cells at sites of inflammation, L-selectin is important in another aspect of leukocyte trafficking. It was originally identified as a homing receptor on lymphocytes when it was shown that L-selectin positive lymphocytes preferentially recirculate or traffic through the peripheral and mesenteric lymph nodes and that L-selectin mediates the binding of lymphocytes to the lymphatic high endothelial venules which are specialized microvessels for lymphocyte emigration. Since L-selectin is expressed by several subsets of leukocytes, mechanisms must exist that regulate L-selectin mediated interactions in a cell-type-specific manner. For example, the restricted recirculation of L-selectin positive lymphocytes through lymphoid tissue, with the exclusion of L-selectin positive neutrophils and other leukocytes, may be dependent upon the expression of different L-selectin ligands by endothelia of that organ. These ligands may differ from each other only in having distinctive modifications, such as sulfation, of a common carbohydrate structure.[67] In addition lineage-specific chemoattractants may lead to the selective recruitment of a subset of naive lymphocytes to HEV, and recruitment of neutrophils to vascular endothelium at sites of inflammation.[2,68]

THE ROLE OF SELECTINS IN DISEASE

Selectins are likely to contribute to many disease states where overzealous leukocyte adhesion and emigration lead to tissue damage. Assessment of selectin function in a variety of animal models of inflammation in vivo have contributed significantly to our understanding of the role of selectins in disease. Only a few of these models will be mentioned. A key role for P- and E-selectin in the development of lung tissue injury has been reported: Cobra-Venom-Factor (CVF), which activates complement, induces an acute inflammatory response (within 30 minutes) in lung and was used to assay P-selectin function in inflammation. Antibodies to P-selectin led to a reduction in permeability and hemorrhage.[69] Antibodies to E-selectin conferred protection against IgG immune complex-induced lung injury and was also effective in reducing dermal vascular injury.[70] Pronounced results have been achieved for all three selectins in several models of ischemia-reperfusion injury. L-selectin antibodies reduced ischemia/reperfusion-induced leukocyte adherence and emigration and albumin leakage in rat mesenteric venules[71] and it attenuated neutrophil accumulation and endothelial dysfunction in a feline model of ischemia/reperfusion.[72] P-selectin antibodies that inhibit leukocyte-endothelial interactions attenuated reperfusion injury in a rabbit ear model[73] and in feline myocardium.[74] E-selectin antibodies administered prior to ischemia and reperfusion of the hind limb of rats showed dramatic decrease in neutrophil entry into the lungs and vascular damage in the lung.[75] Selectins have also been implicated in cancer metastasis. Platelet interaction with cancer cells may aid in blood-borne metastasis and tumor cell adhesion to vascular endothelium may be a prerequisite to tumor extravasation and metastasis. E-selectin mediated adhesion of colon carcinoma cells to cytokine activated vascular endothelium[76] while P-selectin mediated the interaction of activated platelets with small cell lung carcinoma and neuroblastoma cells.[77]

Since selectins are involved in leukocyte recruitment to sites of both acute and chronic inflammation, and may be involved in a wide variety of human diseases including cancer metastasis, blocking their function may be therapeutically advantageous in some circumstances. Approaches to developing inhibitory compounds will rely on a better understanding of selectin-carbohydrate interactions and on more precise information on peptides directly involved in ligand binding. On the basis of such information monoclonal antibodies, peptides and novel carbohydrate based compounds could be developed to block selectin function. The advantage of blocking selectin function rather than the function of integrins or Ig-superfamily molecules is that expression of selectins is restricted to the vasculature, therefore precluding potential side effects that could occur by blocking more widely expressed adhesion molecules. In addition, blocking selectin function would block leukocyte rolling, which is a prerequisite to leukocyte adhesion and emigration and is the earliest known step in the inflammatory process.

REFERENCES

1. Butcher EC. Leukocyte-endothelial cell recognition: three (or more) steps to specificity and diversity. Cell 1991; 67:1033-6.
2. Springer TA. Traffic signals for lymphocyte recirculation and leukocyte emigration: the multistep paradigm. Cell 1994; 76:301-14.
3. Bevilacqua MP, Nelson RM. Selectins. J Clin Invest 1993; 91:379-87.
4. Wagner DD. The Weibel-Palade body: the storage granule for von Willebrand factor and P-selectin. Thromb Haem 1993; 70:105-10.
5. McEver RP. Role of selectins in leukocyte adhesion to platelets and endothelium. Annals of the NY Acad of Sci 1994; 714:185-9.
6. Hattori R, Hamilton KK, Fugate RD et al. Stimulated secretion of endothelial von Wilebrand factor is accompanied by rapid redistribution to the cell surface of the intracellular granule membrane protein GMP-140. J Biol Chem 1989; 264:7768-71.
7. Subramanium M, Koedam JA, Wagner DD. Divergent fates of P- and E-selectins after their expression on the plasma membrane. Mol Biol Cell 1993; 4:791-801.
8. Patel KD, Zimmerman GA, Prescott SM et al. Oxygen radicals induce human endothelial cells to express GMP-140 and bind neutrophils. J Cell Biol 1991; 112:749-759.

9. Sanders WE, Wilson RW, Ballantyne CM et al. Molecular cloning and analysis of in vivo expression of murine P-selectin. Blood 1992; 80:795-800.
10. Weller A, Isenmann S, Vestweber D. Cloning of the mouse endothelial selectins: Expression of both E- and P-selectin is inducible by tumor necrosis factor α. J Biol Chem 1992; 267:15176-15183.
11. Hahne M, Jager U, Isenmann S, et al. Five tumor necrosis factor-inducible cell adhesion mechnisms on the surface of mouse endothelioma cells mediate the binding of leukocytes. J Cell Biol 1993; 121:655-664.
12. Bevilacqua, MP, Pober JS, Mendrick DL et al. Identification of an inducible endothelial-leukocyte adhesion molecule. Proc Natl Acad Sci USA 1987; 84:9238-9242.
13. Kuijpers TW, Raleigh M, Kavanagh T et al. Cytokine-activated endothelial cells internalize E-selectin into a lysosomal compartment of vesiculotubular shape: a tubulin-driven process. J Immunol 1994; 152:5060-5069.
14. Picker LJ, Warnock RA. Burns AR et al. The neutrophil selectin LECAM-1 presents carbohydrate ligands to the vascular selectins ELAM-1 and GMP-140. Cell 1991; 66:921-933.
15. Lasky LA. Selectins: interpreters of cell-specific carbohydrate information during inflammation. Science 1992; 258:964-969.
16. Watson SR, Imai Y, Fennie C et al. The complement binding-like domains of the murine homing receptor facilitate lectin activity. J Cell Biol 1991; 115:235-243.
17. Disdier M, Morrissey JH, Fugate RD, et al. Cytoplasmic domain of P-selectin (CD62) contains the signal for sorting into the regulated secretory pathway. Mol Biol Cell 1992; 3:309-21.
18. Mayadas T, Johnson RC, Rayburn H et al. Leukocyte rolling and extravasation are severely compromised in P-selectin-deficient mice. Cell 1993; 74:541-54.
19. Crovello CS, Furie BC, Furie B. Rapid phosphorylation and selective dephosphorylation of P-selectin accompanies platelet activation. J Biol Chem 1993; 268:14590-3.
20. Fujimoto T, McEver RP The cytoplasmic domain of P-selectin is phosphorylated on serine and threonine residues. Blood 1993; 82:1758-66.
21. Ushiyama S, Laue TM, Moore KL et al. Structural and functional characterization of monomeric soluble P-selectin and comparison with membrane P-selectin. J Biol Chem 1993; 268:15229-37.
22. Kansas GS, Ley K, Munro JM et al. Regulation of leukocyte rolling and adhesion to HEV through the cytoplasmic domain of L-selectin. J Exp Med 1993; 177:833-8.
23. Firestein GS, Bullough DA, Erion MD et al. Inhibition of neutrophil adhesion by adenosine and an adenosine kinase inhibitor—the role of selectins. J Immunol 1995; 154:326-34.

24. Fukuda M, Spooncer E, Oates JE. Structure and sialylated fucosyl lactosaminoglycan isolated from human granulocytes. J Biol Chem 1984; 259:10925-35.
25. Walz G, Aruffo A, Kolanus W et al. Recognition by ELAM-1 of the sialyl-Lex determinant on myeloid and tumor cells. Science 1990; 250:1132-5.
26. Munro JM, Lo SK, Corless C et al. Expression of sialyl-Lewis x, an E-selectin ligand, in inflammation, immune processes, and lymphoid tissues. Am J Pathol 1992; 141:1397-1408.
27. Ohmori, K, Yoneda T, Ishihara G et al. Sialyl SSEA-1 antigen as a carbohydrate marker on human natural killer cells and immature lymphoid cells. Blood 1989; 74:255.
28. Magnani JL, Nilsson B, Brockhaus M et al. A monoclonal antibody-defined antigen associated with gastrointestinal cancer is a ganglioside containing sialylated lacto-N-fucopentaose II. J Biol Chem 1982; 23:14365-9.
29. Fukushima K, Hirota M, Terasaki P et al. Characterization of sialosylated Lewis x as a new tumor-associated antigen. Cancer Res 1984; 44:5279-85.
30. Aruffo A, Dietsch MT, Wan H et al. Granule membrane protein 140 (GMP140) binds to carcinomas and carcinoma-derived cell lines. Proc Natl Acad Sci 1992; 89:2292-6.
31. Aruffo A, Kolanus W, Walz G et al. CD62/P-selectin recognition of myeloid and tumor cell sulfatides. Cell 1991; 67:35-44.
32. Varki A. Selectin ligands. Proc Natl Acad Sci USA 1994; 91:7390-7.
33. Imai Y, Singer MS, Fennie C et al. Identification of a carbohydrate-based endothelial ligand for a lymphocyte homing receptor. J Cell Biol 1991; 113:1213-21.
34. Lasky LA, Singer MS, Dowbenko D et al. An endothelial ligand for L-selectin is a novel mucin-like molecule. Cell 1992; 69:927-38.
35. Baumhueter S, Singer MS, Henzel W et al. Binding of L-selectin to the vascular sialomucin CD34. Science 1993; 262:436-8.
36. Spertini O, Luscinskas FW. Kansas GS et al. Leukocyte adhesion molecule (LAM-1, L-selectin) interacts with an inducible endothelial ligand to support leukocyte adhesion. J Immunol 1991; 147:2565-73.
37. Sako D, Chang X-J, Barone KM et al. Expression cloning of a functional glycoprotein ligand for P-selectin. Cell 1993; 75:1179-86.
38. Abbassi O, Kishimoto TK, McIntire LV et al. E-selectin supports neutrophil rolling in vitro under conditions of flow. J Clin Invest 1993; 92:2719-30.
39. Lenter M, Levinovitz A, Isenmann S et al. Monospecific and common glycoprotein ligands for E- and P-selectin on myeloid cells. J Cell Biol 1994; 125:471-81.
40. Levinovitz A, Muhlhoff J, Isenmann S et al. Identification of a glycoprotein ligand for E-selectin on mouse myeloid cells. J Cell Biol 1993; 121:449-59.

41. Walcheck B, Watts G, Jutila MA. Bovine gamma/delta T cells bind E-selectin via a novel glycoprotein receptor: first characterization of a lymphocyte/E-selectin interaction in an animal model. J Exp Med 1993; 178:853-63.
42. Anderson DC, Springer TA. Leukocyte adhesion deficiency: an inherited defect in the Mac-1, LFA-1 and p150,95 glycoproteins. Ann Rev Med 1987; 38:175-94.
43. Etzioni A, Frydman M, Pollack S, et al. Brief report: recurrent severe infections caused by a novel leukocyte adhesion deficiency. New Engl J Med 1992; 327:1789-92.
44. Frydman M, Etzioni A, Eidlitz-Markus T et al. Rambam Hasharon syndrome of mental retardation, short-limbed dwarfism, defective neutrophil chemotaxis and Bombay phenotype. Am. J Med Genet 1993; 44:297-302.
45. von Andrian UH, Berger EM, Ramezani L et al. In vivo behavior of neutrophils from two patients with distinct inherted leukocyte adhesion deficiency syndromes. J Clin Invest 1993; 91:2893-7.
46. Thomas KR, Capecchi MR. Site-directed mutagenesis by gene targeting in mouse embryo-derived stem cells. Cell 1987; 51:503-12.
47. Mansour SL, Thomas KR, Capecchi MR. Disruption of the proto-oncogene *int-2* in mouse embryo-derived stem cells: a general strategy for targeting mutations to non-selectable genes. Nature 1988; 336:348-52.
48. Smith CW, Kishimoto TK, Abbass O et al. Chemotactic factors regulate lectin adhesion molecule 1 (LECAM-1)-dependent neutrophil adhesion to cytokine-stimulated endothelial cells in vitro. J Clin Invest 1991; 87:609-18.
49. Spertini O, Luscinskas FW, Gimbrone Jr MA et al. Monocyte attachment to activated human vascular endothelium in vitro is mediated by leukocyte adhesion molecule-1 (L-selectin) under nonstatic conditions. J Exp Med 1992; 175:1789-92.
50. Brady HR, Spertini O, Jimenez W et al. Neutrophils, monocytes, and lymphocytes bind to cytokine-activated kidney glomerular endothelial cells through L-selectin (LAM-1) in vitro. J Immunol 1992; 149:2437-44.
51. Ley K, Gaehtgens P, Fennie C et al. Lectin-like cell adhesion molecule 1 mediates leukocyte rolling in mesenteric venules in vivo. Blood 1991; 77:2553-5.
52. von Andrian UH, Chambers JD, McEvoy LM et al. Two-step model of leukocyte-endothelial cell interaction in inflammation: distinct roles for LECAM-1 and the leukocyte beta-2 integrins in vivo. Proc Natl Acad Sci USA 1991; 88:7538-42.
53. Arbones ML, Ord DC, Ley K et al. Lymphocyte homing and leukocyte rolling and migration are impaired in L-selectin-deficient mice. Immunity 1994; 1:247-60.

54. Ley K, Bullard DC, Arbones ML et al. Sequential contribution of L- and P-selectin to leukocyte rolling in vivo. J Exp Med 1995; 181:669-75.
55. Dore M, Hawkins HK, Entman ML et al. Production of monoclonal antibody against canine GMP-140 (P-selectin) and studies of its vascular distribution in canine tissues. Vet Pathol 1993; 30:213-22.
56. McEver RP, Beckstead JH, Moore KL et al. GMP-140, a platelet a-granule membrane protein, is also synthesized by vascular endothelial cells and is localized in Weibel-Palade bodies. J Clin Invest 1989; 84:92-9.
57. Olofsson AM, Arfors KE, Ramezani L et al. E-selectin mediates leukocyte rolling in interleukin-1-treated rabbit mesentery venules. Blood 1994; 84:2749-58.
58. Lorant DE, Patel KD, McIntyre TM et al. Coexpression of GMP-140 and PAF by endothelium stimulated by histamine or thrombin: a juxtacrine system for adhesion and activation of neutrophils. J Cell Biol 1991; 115:223-34.
59. Lorant DE, Topham MK, Whatley RE et al. Inflammatory roles of P-selectin. J Clin Invest 1993; 92:559-70.
60. Waddell TK, Fialkow L, Chan CK et al. Potentiation of the oxidative burst of human neutrophils–a signaling role for L-selectin. J Biol Chem 1994; 269:18485-91.
61. Watson S, Fennie C, Lasky LA. Neutrophil influx into an inflammatory site is inhibited by a soluble homing receptor-IgG chimaera. Nature 1991; 349:164-7.
62. Bosse R, Vestweber D. Only simultaneous blocking of the L- and P-selectin completely inhibits neutrophil migration into mouse peritoneum. Eur J Immunol 1994; 24:3019-24.
63. Labow MA, Norton CR, Rumberger JM et al. Characterization of E-selectin-deficient mice: demonstration of overlapping function of the endothelial selectins. Immunity 1994; 1:709-20.
64. Larsen E, Celi A, Gilbert GE et al. PADGEM protein: a receptor that mediates the interaction of activated platelets with neutrophils and monocytes. Cell 1989; 59:305-312.
65. Hamburger SA, McEver RP. GMP-140 mediates adhesion of stimulated platelets to neutrophils. Blood 1990; 75:550-4.
66. Palabrica T, Lobb R, Furie BC et al. Leukocyte accumulation which promotes fibrin deposition is mediated in vivo by P-selectin (CD62) on adherent platelets. Nature 1992; 359:848-51.
67. Rosen SD, Bertozzi CR. The selectins and their ligands. Curr Opin Cell Biol 1994; 6:663-73.
68. Spertini O, Kansas GS, Munro JM et al. Regulation of leukocyte migration by activation of the leukocyte adhesion molecule-1 (LAM-1) selectin. Nature 1991; 349:691-4

69. Mulligan MS, Polley MJ, Bayer RJ et al. Neutrophil-dependent acute lung injury: requirement for P-selectin (GMP-140). J Clin Invest 1992; 90:1600-7.
70. Mulligan MS, Varani J, Dame MK et al. Role of endothelial-leukocyte adhesion molecule 1 (ELAM-1) in neutrophil-mediated lung injury in rats. J Clin Invest 1991; 88:1396-1406.
71. Kurose I, Anderson DC, Miyasaka M et al. Molecular determinants of reperfusion-induced leukocyte adhesion and vascular protein leakage. Circulation Res 1994; 74:336-43.
72. Ma X, Weyrich AS, Lefer DJ et al. Monoclonal antibody to L-selectin attenuates neutrophil accumulation and protects ischemic reperfused cat myocardium. Circulation 1993; 88:649-58.
73. Winn RK, Liggitt D, Vedder NB. Anti-P-Selectin monoclonal antibody attenuates reperfusion injury to the rabbit ear. J Clin Invest 1993; 92:2042-7.
74. Weyrich AS, Ma X, Lefer DJ et al. In vivo neutralization of P-selectin protects feline heart and endothelium in myocardial ischemia and reperfusion injury. J Clin Invest 1993; 91:2620-9.
75. Seekamp A, Till GO, Mulligan MS et al. Role of selectins in local and remote tissue injury following ischemia and reperfusion. Am J Pathol 1994; 144:592-8.
76. Rice GE, Bevilacqua MP. An inducible endothelial cell surface glycoprotein mediates melanoma adhesion. Science 1989; 246:1303-6.
77. Stone JP, Wagner DD. P-selectin mediates adhesion of platelets to neuroblastoma and small cell lung cancer. J Clin Invest 1993; 92:804-13.
78. Yamada S, Mayadas TN, Yuan F, Wagner DD, Hynes RO, Melder RJ, Jain RK. Rolling in P-selectin-deficient mice is reduced but not eliminated in the dorsal skin. Blood 1995; in press.
79. Johnson RC, Mayadas TN, Frenette PS et al. Blood cell dynamics in P-selectin-deficient mice. Blood 1995; 86:1106-1114.
80. Steegmaler M, Levinovitz A, Isenmann S, Borges E, Lenter M, Kocher HP, Kleuser B, Vestweber D. The E-selectin-ligand ESL-1 is a variant of a receptor for fibroblast growth factor. Nature 1995; 373:615-620.

CHAPTER 7

IN VITRO MODELS OF INVASION

John Nip and Pnina Brodt

Tumor cell invasion is a key event in the complex process of cancer metastasis. The metastatic potential of tumor cells depends on their ability to adhere to, degrade, and migrate through extracellular matrix (ECM) barriers such as stroma and basement membranes.[1] Several families of proteinases have been implicated in ECM degradation and tumor invasion. Three families have been grouped on the basis of the amino acid residues critical for the catalytic activity and include the serine, cysteine and aspartic acid proteinases and the fourth the metalloproteinases on the basis of their common dependence on a metal ion, generally Zn^{2+}, for catalytic activity (for review see ref. 2). In order to better understand the factors mediating invasion, various in vitro models have been developed which attempt to isolate and mimic discrete steps within the metastatic process. These models, which range in complexity from the simple matrix degradation models to the more elaborate organ culture models, have provided much of the information now available on the proteolytic processes and mediators involved in cancer invasion and metastasis. They are summarized briefly in this chapter.

As suggested by Mareel et al, the in vitro assays used to study invasion can essentially be divided into two groups based on whether the substratum confronted by the invading cells is two- or three-dimensional.[3]

TWO-DIMENSIONAL SYSTEMS

One of the first two-dimensional model systems to be used was developed by Santesson who examined the interactions of

Cell Adhesion and Invasion in Cancer Metastasis, edited by Pnina Brodt.

malignant mouse mammary tumor explants with chick embryo heart organ cultures.[4,5] He found that cells growing out from the more malignant mouse tumors intermingled with the cells derived from the heart explants whereas the less malignant tumors formed a distinctive border with the normal tissue. Other groups[6,7] used similar methods in combination with time lapse cinematography to obtain more qualitative information on the invasion of a variety of tumors.

Another type of two-dimensional in vitro model used is the cell-monolayer invasion system. In this model, a monolayer, usually of non-malignant cells is maintained on culture dishes. Tumor cells in suspension are added on the monolayer and the invasion is monitored and quantitated using phase contrast or electron microscopy[3,8-10] which can be combined with time-lapse video microscopy[11] (Fig. 7.1).

In addition to invasion assays involving cell-cell interactions, matrix proteins have also been used in two-dimensional assays. Typically, tumor cells were seeded onto a protein substratum consisting of single basement membrane proteins such as type IV collagen[12] or on more complex matrices such as human amnion[13] (for review see ref. 14). As tumor invasion generally involves proteolytic degradation of the substratum, it can be monitored by measuring the release of proteolytic fragments from an isotope-labeled substrate into the assay medium (Fig. 7.1b). The isotopes used in these assays have included ^{45}Ca, which was used to metabolically label mouse calvaria cells;[15] or [^{3}H]leucine, [^{3}H]proline, and [^{35}S]methionine which were used to metabolically label the subendothelial matrix.[16] Some substrata, such as bovine lens capsule basement membrane have been surface-labeled with ^{125}Iodine.[17]

THREE-DIMENSIONAL SYSTEMS

The two-compartment chamber model is one example of a three-dimensional invasion system. The original system developed by Hart and Fidler[18] consisted of a two-compartment invasion chamber prepared from latex cylinders separated by a chick chorioallantoic membrane. In this assay, radiolabeled tumor cells were added to the upper compartment, allowed to penetrate the membrane and were then trapped in a photographic sponge. Quantitation of the radioactivity retained in the sponge provided an indirect measure of cell invasiveness and showed a positive correlation

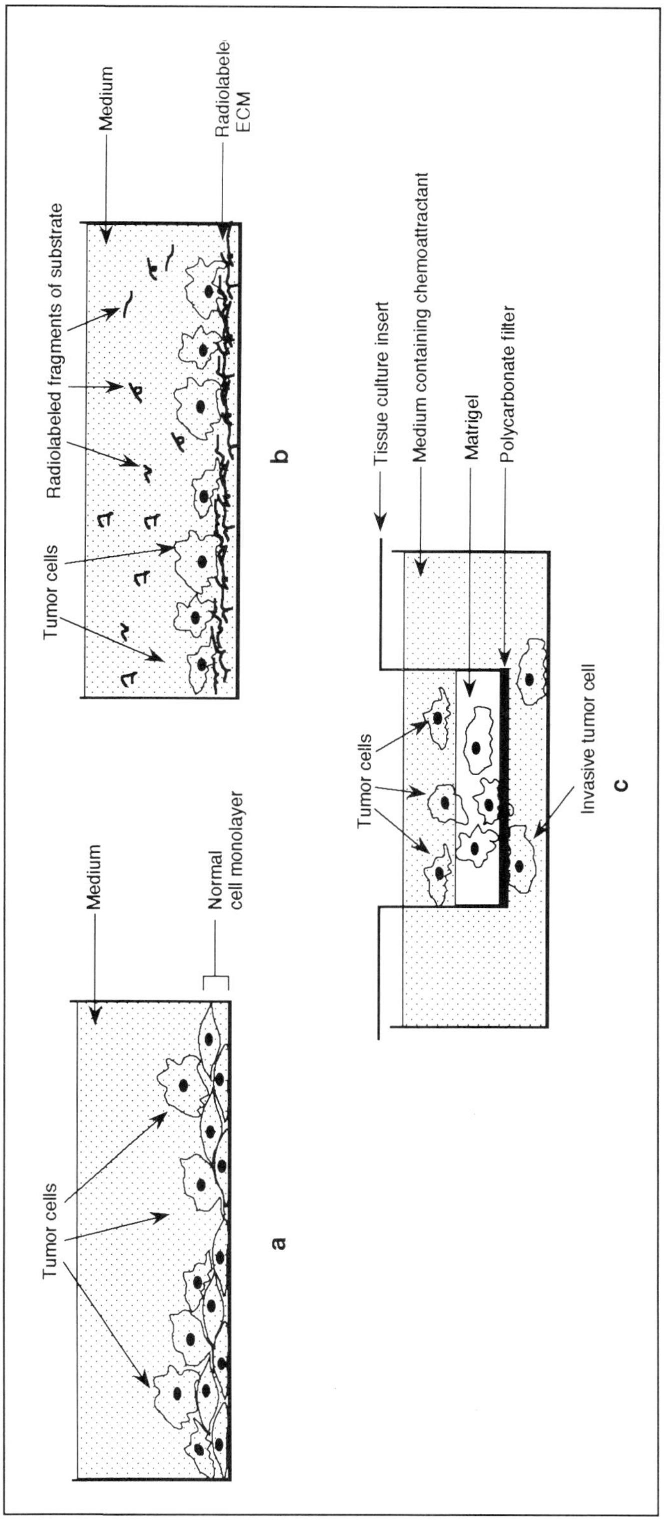

Fig. 7.1. In vitro assays of ECM degradation and invasion. A diagrammatic representation of tumor cells (a) invading through a cell monolayer;[8-10] (b) degrading radiolabeled ECM;[16] and (c) invading through Matrigel.[19]

between invasion and the metastatic potential of the cells in vivo. This system lends itself to a qualitative analysis when combined with histology.

A modification of the two-compartment chamber model is the widely used Matrigel invasion assay. This assay, which measures the invasion of cells through a basement membrane-like barrier, was originally described by Albini et al.[19] It is similar to the above mentioned two-compartment systems except that in place of a biological membrane, a polycarbonate membrane filter is used which is precoated with a reconstituted basement membrane gel derived from the conditioned medium of the Engelbreth-Holm-Swarm mouse sarcoma line (Matrigel). The tumor cells are added to the upper chamber and allowed to penetrate the Matrigel. The invading cells can then transverse the 8 μM-pore filters and depending on the cells and assay conditions, can either be harvested from the lower chamber or may attach and spread on the lower surface of the filter where they can be stained and enumerated (Fig. 7.1c). In place of Matrigel-coated filters, placental amnion membranes have also been used as well as filters coated with isolated extracellular matrix proteins such as collagens or fibronectin.[20]

Another type of three-dimensional confrontation model are the organ cultures (reviewed in ref. 20) which measure the ability of tumor cells to invade cultured organ fragments. For these assays, a variety of organs have been used including chick embryo heart fragments, rat omentum, lung, and bladder and human prostate. Tumor cell invasion into the organ fragments is normally monitored using histological sections of the tissue. Alternatively, invasion can be quantitated using isotope-labeled tumor cells,[21] but this method cannot easily distinguish between adherent and invasive cells. The major advantage of these systems has been their ability to more accurately mimic the conditions encountered by invading tumor cells in vivo.

In addition, three-dimensional confrontation models involving cell or tissue aggregates have been used (reviewed in refs. 5, 22). These involve the use of three-dimensional cell clusters which are allowed to interact with aggregates of normal tissue in rotation cultures. Invasion is subsequently measured using microscopic analysis of tissue sections or by assessing the depth of penetration of radiolabeled tumor cell aggregates into normal cell aggregates. One of the earliest examples for the usage of this assay system has been a study by Moscona[23] using rotation cultures of chick embryo liver cells and pigmented murine melanoma cells. Although not originally designed to measure invasion, this study did demonstrate that the melanoma cells invaded the liver cell aggregates. Mareel et al[3] later used an organ culture system consisting of rounded fragments of chick embryo heart incubated with aggregates of mouse sarcoma cells in rotation cultures. The invasion of heart tissue in these studies was assessed by histology.

Adaptations and modifications of these in vitro invasion systems are still being used today for the study of invasion mechanisms and their role in metastasis. With the advent of new technologies, such as computer-assisted imaging systems, these analyses are becoming more quantitative. Methods such as confocal microscopy have made it unnecessary to use histology for analysis of tissue invasion, as optical sectioning has replaced physical disruption of the tissue.[24] These powerful tools will provide more precise and analytical methods for the study of cell invasion.

References

1. Liotta L, Rao C, Barsky S. Tumor invasion and the extracellular matrix. Lab Invest 1983; 49:636-49.
2. Mignatti P, Rifkin DB. Biology and biochemistry of proteinases in tumor invasion. Physiological Reviews 1993; 73(1):161-87.
3. Mareel MM, Van Roy FM, Messiaen LM et al. Qualitative and quantitative analysis of tumour invasion in vivo and in vitro. J Cell Sci Suppl 1987; 8:141-63.
4. Santesson L. Characteristics of epithelial mouse tumor cells in vitro and tumor structures in vivo. A comparative study. Acta Pathol Microbiol Scand Suppl 1935; 24:1-237.
5. Easty GC, Easty DM. In vivo and in vitro models of invasion. In: Invasion experimental and clinical implications. Oxford: Oxford University Press 1984:24-62.
6. Abercrombie M, Heaysman JEM. Invasive behaviour between sarcoma and fibroblast populations in cell culture. J Natl Cancer Inst 1976; 56:561-70.
7. Easty GC, Haemmerli G, Easty DM et al. Interactions between normal epithelial and squamous carcinoma cells in monolayer culture. Cancer Res 1982; 42:4248-55.
8. Horai T, Imamura F, Nakae T et al. Interactions of lung cancer cells with the human mesothelium cell monolayer: an in vitro model for cancer invasion. Japanese J Clin Oncology 1992; 22(2):67-72.
9. Akedo H, Shinkai K, Mukai M. et al. Interaction of rat ascites hepatoma cells with cultured mesothelial cell layer: a model for tumor invasion. Cancer Res 1986; 46:2416-22.

10. Roos E, van der Pavert IV, Middelkoop OP. Infiltration of tumour cells into cultures of isolated hepatocytes. Cancer Res 1981; 46:2416-22.
11. Verchueren H, De Baetselier P, Bereiter-Hahn J. Dynamic morphology of metastatic mouse T-lymphoma cells invading through monolayers of 10T1/2 cells. Cell Motility and the Cytoskeleton 1991; 20(3):203-14.
12. Murray JC, Garbisa S, Liotta L. The role of tumor cell-basement membrane interactions in the metastatic process. Oncol 1980; 4:169-73.
13. Russo RG, Thorgeirsson U, Liotta L. In vitro quantitative assay of invasion using human amnion. In: Liotta LA, Hart IR, eds. Tumor invasion and metastasis. Boston: Martinus Nijhoff Publishers 1982; 173-187.
14. Mareel MM. Invasion in vitro. Methods of analysis. Cancer Metastasis Rev 1983; 2:201-18.
15. Tsao SW, Burman JF, Easty DM et al. Some mechanisms of local bone destruction by squamous carcinomas of the head and neck. Br J Cancer 1981; 43:392-401.
16. Kramer RH, Vogel KG, Nicolson GL. Solubilization and degradation of subendothelial matrix glycoproteins and proteoglycans by metastatic tumor cells. J Biol Chem 1982; 257:2678-86.
17. Starkey JR, Hosick HL, Stanford DR et al. Interaction of metastatic tumor cells with bovine lens capsule basement membrane. Cancer Res 1984; 44(4):1585-94.
18. Hart IR, Fidler IJ. An in vitro quantitative assay for tumor cell invasion. Cancer Res 1978; 38:3218-24.
19. Albini A, Iwamoto Y, Kleinman HK et al. An in vitro quantitative assay for tumor cell invasion. Cancer Res 1987; 47:3239-45.
20. Easty DM, Easty GC. An in vitro model for studying cell invasiveness. In: Organ Culture in Biomedical Research. Cambridge: Cambridge University Press 1976; 379-93.
21. Waller CA, Braun M, Schirrmacher V. Quantitative analysis of cancer invasion in vitro: comparison of two new assays and of tumour sublines with different metastatic capacity. Clin Expl Metast 1986; 4:73-89.
22. Gershman H. Three dimensional models for the study of invasion and metastasis. In: Liotta LA, Hart IR, eds. Tumor Invasion and Metastasis. Boston: Martinus Nijhoff Publishers 1982; *i*:231-250.
23. Moscona A. Rotation-mediated histogenetic aggregation of dissociated cells. Exp Cell Res 1961; 22:455-75.
24. Petroll MM, Jester JV, Cavanagh HD. In vivo confocal imaging: general principles and applications. Scanning 1994; 16(3)131-49.

CHAPTER 8

Structure Function Requirements of Urinary Plasminogen Activator (uPA) and its Receptor

Shafaat A. Rabbani

INTRODUCTION

A critical event in the fibrinolytic process is the conversion of inactive proenzyme plasminogen, a plasma protein ubiquitously distributed in the body, to its active moiety plasmin by plasminogen activators (PAs).[1] The two main members of this family of serine proteases are tissue plasminogen activator (tPA) and urinary plasminogen activator (uPA), which are the products of discrete genes.[2,3] Although these PAs share 40% sequence homology, they are immunologically distinct and only uPA can bind to its specific receptor, uPAR.[4] Both are produced by several normal cell types and may be secreted simultaneously by the same cells. The primary role of tPA is connected to intravascular thrombolysis whereas uPA plays a dominant role in cell migration, during diverse processes such as wound healing, angiogenesis and tumor metastasis.[5-8] The proteolytic activity of plasminogen activators is regulated by inhibitors of both plasmin and PA. PA inhibitors belong to the serine proteinase inhibitor (serpin) superfamily and are classified as

Cell Adhesion and Invasion in Cancer Metastasis, edited by Pnina Brodt.

arginine-serpins. The inhibition of enzyme activity is accomplished by formation of 1:1 covalent complexes between enzyme and inhibitor. Two major functionally active PA inhibitors (PAIs) have been characterized, including the endothelial-type PAI (PAI-1) and the placental-type PAI (PAI-2).[9] Plasminogen activators have also been implicated in the pathology of malignant disease. Cells derived from malignant tumors produce PAs more often and at higher levels than their normal counterparts.[10] In the majority of studies the PA implicated in malignancy is uPA. Increased secretion of PAs results in decreased tumor cell adhesion to substratum and increased invasion into the surrounding tissues.[11] A strong correlation therefore exists between the secretion of uPA and metastatic ability of tumor cells.[11]

UROKINASE

Urokinase is synthesized as a single chain propeptide, (Pro-uPA), which then undergoes cleavage to form a two-chain serine protease connected by a disulfide bridge which is known as High Molecular Weight uPA (HMW uPA).[12] The 24 kDa A, or light chain, is formed from the NH_2-terminal end of the single chain and contains one "kringle" structure in its NH_2-terminal region and a cysteine rich "growth factor domain" (GFD) which is structurally similar to the receptor-binding domain of EGF.[13] The His-Asp-Ser containing, enzymatic domain resides in the 30 kDa B, or heavy chain. Cleavage of the A-chain may occur, causing release of the first 135 NH_2-terminal residues which is known as the Amino Terminal Fragment (ATF 1-135), containing the GFD, and resulting in the Low Molecular Weight form of uPA which retains plasminogen activator function (LMW uPA). tPA is also secreted as a single chain form containing a disulfide bond and also contains a "growth factor" module. This chain also undergoes cleavage to form an active two chain plasminogen activator. The A chain of tPA shares a high degree of sequence homology with the A chain of uPA but is larger in size. Unlike the A chain of uPA it does not appear to undergo a second cleavage step to release an entity containing the growth factor domain (Fig. 8.1).

cDNAs encoding uPA have been isolated from human, baboon, mouse and rat tissues and their deduced amino acid sequence determined.[2,14-17] uPAs from these species share a high degree of nucleotide and amino acid sequence homology, however significant differences have been noted as well.[14-17] These cDNAs have a long 3' non-coding region with two potential polyadenylation sites. However, Northern blot analysis of multiple tissues revealed the presence of a single uPA mRNA in multiple tissues from various species suggesting the predominant usage of the second polyadenylation site. Other features in these cDNAs such as a long poly A^+ tail and palindromic sequence may contribute to the expression and stability of uPA mRNA.

Despite the high sequence homology among uPAs of different species, human uPA was found to be less effective in receptor binding and mitogenic activity when tested in rat osteoblasts as compared to human osteosarcoma cells (SaOS-2).[18] Additionally it has been reported that synthetic mouse uPA fragments do not bind to human U-937 lymphoma cells but do bind to mouse cells.[13] The

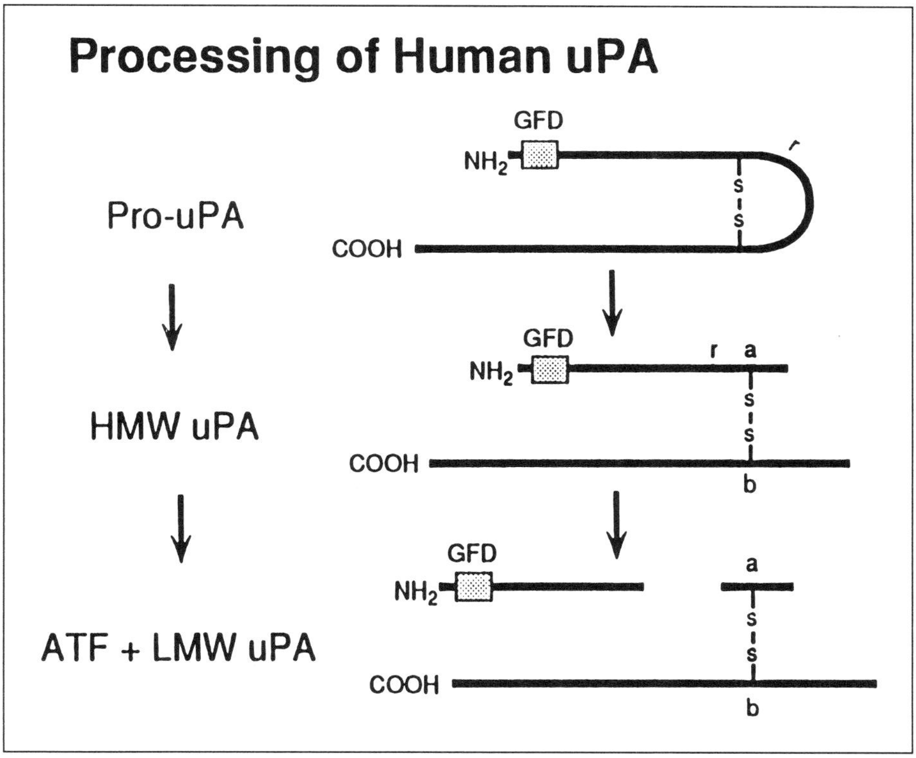

Fig. 8.1. Diagram of the processing of human urokinase (uPA). The single chain Pro-uPA is first cleaved to high molecular weight (HMW) uPA which then undergoes another cleavage to generate low molecular weight (LMW) uPA and an amino terminal fragment (ATF). GFD represents the growth factor domain and "a" and "b" signifies the two chains of uPA linked by disulfide bridge (S-S) bond.

deduced amino acid sequence from rat uPA cDNA recently cloned in our laboratory predicts a 431 amino acid protein which is 69% and 85% homologous with its human and mouse counterparts, respectively.[16,17] However the degree of conservation is lower at the apparent receptor binding region located within the GFD. This region of uPA (residues 13-33) shares partial amino acid sequence homology with EGF (residues 14-33) and in both molecules has been reported to be responsible for binding to their cognate receptors.[13] No cross-reactivity of these ligands to the non-cognate receptor has been observed. Four cysteines are conserved within this receptor binding sequence and the region of highest homology is between residues 13-19, and 14-20 of uPA and EGF respectively both of which are located in the first loop of the GFD (Fig. 8.2). Inasmuch as previous studies with synthetic fragments have shown that both regions are required for optimal receptor binding, residues 13-19 of uPA apparently confer binding specificity whereas residues 20-30 seem to provide the proper conformation to the binding region.[13] There are four amino acid differences in the middle loop between rat and human uPA (Tyr_{22} Ser_{27} Arg_{29} Arg_{30}) and two (Gln_{14} Val_{18}) in the first loop. One of the potentially critical substitutions in the first loop of rat uPA is Val_{18} for Thr_{18} (found in the human). This substitution eliminates the possibility of fucosylation found necessary for the mitogenic effect of human uPA and may alter the conformation of the sequence in the middle loop thus impeding signal transduction in human

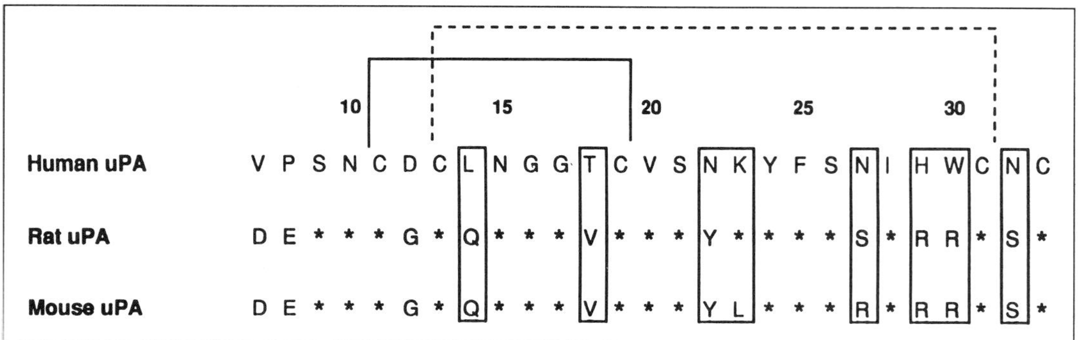

Fig. 8. 2. Comparison of the amino acid sequence of receptor binding domain (13-33) of rat, human and mouse uPA. The amino acid sequence in human and mouse uPA identical with rat are denoted by an asterisk (*). Positions of sequence variability are shown in boxed area. Solid (—) and broken lines (- - -) represent disulfide linkages.

cells.[19,20] This Val_{18} substitution also occurs in mouse uPA[2]. Consequently, this alteration may also contribute to the impaired capacity of mouse uPA to bind to human cells and the diminished reactivity of human uPA for rat osteoblastic cells. Inasmuch as defucosylated human uPA can still bind to its receptor, the four substitutions between rat and human uPA in the middle loop must provide the major contribution to differences in receptor binding between species. One less difference (Lys_{23}) between rat and human uPA and a more conservative substitution at position 27 than that occurring in mouse uPA may be responsible for better cross-reactivity between rat and human uPA and their receptors than between the corresponding mouse and human species.

Other post-translational modifications of uPA include glycosylation and phosphorylation.[20,21] Glycosylation of uPA in the GED of uPA results in increased activation of plasminogen and rapid inactivation by plasma inhibitors.[22] Although phosphorylation of uPA has been shown to cause no difference in plasminogen activation, however, it decreases the sensitivity of uPA to PAI-1, which may indirectly result in increased uPA activity.[21] These post-translational modifications of uPA in its receptor-binding and enzymatic domains may also play a critical role in tumor invasion and metastasis by changing its enzymatic activity and plasma clearance characteristics.

THE DUAL ROLE OF uPA IN SKELETAL METASTASIS

In our previous studies, amino acid sequence and amino acid analysis were consistent with the presence of an ATF of uPA with osteoblastic activity isolated from the conditioned medium of the human prostate carcinoma PC-3 cells.[14] The presence of this NH_2-terminal fragment suggested that carboxyl terminal proteolysis of the expected NH_2-terminal cleavage product had occurred or that processing of uPA occurs somewhat differently in the prostatic cancer cells we examined. The finding of growth factor activity for osteoblastic cells in an NH_2-terminal fragment of uPA isolated from PC-3 cells conditioned medium may be of fundamental importance in understanding the mechanism of development of osteoblastic metastases associated with prostate cancer. In these studies an amino terminal fragment of uPA (ATF 1-143), its growth factor domain (GFD-4-43) both of which are fucosylated at Thr_{18} caused a dose dependent increase in $[^3H]$ thymidine incorporation in SaOS-2 human osteosarcoma cells. Removal of fucose at Thr_{18} either by chemical means or by expression of uPA in *E. coli* where it is not fucosylated resulted in the loss of its mitogenic effects suggesting that the carbohydrate moiety plays a critical role in inducing these growth factor-like effects in osteoblasts. Furthermore the structurally related tPA, had no mitogenic effects in these cells.[19]

The presence of growth factor activity and plasminogen activator activity in two discrete molecular domains of uPA may have important implications for the role of uPA in malignancy in general and in mediating osteoblastic skeletal lesions in particular. Thus the proteolytic activity of uPA secreted by malignant cells metastatic to bone may well be involved in breakdown of the extracellular skeletal matrix by mechanisms similar to basement membrane and connective tissues digestion and the production of lytic osseous lesions. This activity may indeed predominate when the high molecular weight form of uPA is intact or if the NH_2-terminal fragment is rapidly degraded. However, overproduction or persistence of the NH_2-terminal moiety may provide a stimulus for osteoblastic proliferation which could then become the dominant phenomenon. This might explain the predominance of lytic lesions in association with many skeletal metastases but also the infrequent but definite potential for osteoblastic lesions associated with a variety of tumors. The propensity for production of osteoblastic metastases exhibited by tumors such as prostatic carcinoma may be a function of its capacity to more efficiently generate osteoblast-stimulating NH_2-terminal fragments and/or to release plasminogen activator inhibitors to neutralize the proteolytic activity.

THE UROKINASE RECEPTOR

In the complex process of tumor invasion and metastasis increasing evidence has implicated the uPA receptor (uPAR). In this process the saturation of uPAR on the tumor cell surface membrane serves to focus the pericellular proteolytic activity of uPA hence the direct correlation demonstrated between matrix and matrix degradation and uPAR occupancy.[23] A high affinity cell surface uPAR was first reported to be present on the human lymphoma cells U-937.[24] This uPAR subsequently identified in diverse cell types can bind single chain uPA and HMW uPA and requires the presence of non-enzymatic amino terminal fragment

(ATF 1-135) containing the growth factor domain.[25] The minimum sequence requirement for receptor binding has been further localized within amino acids 12-32 by comparing the binding characteristics of synthetic fragments corresponding to this domain. This part of uPA shares a strong amino acid homology with the receptor binding region of epidermal growth factor (EGF).[26] However, despite this sequence homology, binding of uPA is quite specific since no other related ligands have been reported to bind to uPAR. Single chain uPA can be activated to HMW uPA even when bound to its receptor and receptor-bound HMW uPA maintains its ability to catalyze plasmin formation from plasminogen. As the receptor was localized to regions of contact between the cell membrane and the extracellular matrix (ECM) namely focal contacts, this serves to focus the proteolytic activity to cell ECM boundary providing a mechanism for coordination of cell adhesion, proteolysis and motility.[27]

Several lines of evidence suggest that membrane anchored uPAR is involved in signal transduction mechanisms.[28] Thus, HMW uPA but not LMW uPA lacking the ATF has been reported to have proliferative effects in selected cell types, such as an epidermoid tumor cell line.[29] Consequently, it has been postulated that binding of uPA to its receptor can induce cellular proliferation via an autocrine uPA-uPAR loop. (To date, no proliferative effects of tPA have been reported.) Unlike other receptors, uPAR after binding of its specific ligand does not undergo internalization and the receptor-ligand complex remains easily accessible on the cell surface. However, once ECM-associated PAI-1 binds to this complex, the enzyme is rapidly inactivated with rapid removal of the uPA/uPAR-PAI complex which is internalized and degraded in lysosomes.[30,31] This phenomenon is a common mechanism of action for serine protease-serpins system as tPA/PAI-1 complex formation can be inhibited by the uPA/PAI-1 complex, suggesting that they compete for the actions of PAI-1.

Recently cDNAs encoding both human and mouse uPA receptors (uPARs) have been isolated and the gene for human uPAR localized to the long arm of chromosome 19.[32-34] The receptor is a cysteine-rich, highly glycosylated 55-60 kDa protein synthesized as a 22 amino acid signal peptide followed by a 313 amino acid mature peptide with five potential sites for N-linked glycosylation. While most membrane-anchored proteins are linked to the plasma membrane through hydrophobic interactions at the transmembrane region, experimental evidence suggests that uPAR utilizes an alternate mode of attachment to transduce its effects. This is achieved by receptor anchorage via a covalent linkage of the COOH terminus of uPAR to the glycosylated phospholipid phosphatidylinositol.[28] This linkage results in the formation of a strong complex termed the glycosyl-phosphatidylinositol (GPI) anchor.[35,36] This GPI anchor not only provides a mechanism of membrane attachment for uPAR but also plays an important role in eliciting its biological effects, as it results in increased lateral mobility of uPAR within the plasma membrane bilayer and a more efficient recruitment of uPA by the receptor. This process is believed to play an important role in the cascade of cell migration and tumor invasion where uPAR is implicated as a major mediator. Characterization of the uPAR protein revealed that the actual cell surface bound receptor is considerably smaller than that predicted by its cDNA sequence. This is due to post-translational elimination of a hydrophobic region by proteolytic processing at its COOH terminus, an event shared by other members of the GPI-linked proteins. In the case of uPAR, post-translational proteolytic cleavage of a third of the COOH terminal amino acids results in the formation of a GPI anchor at Ser_{282}, Gly_{283} or Ala_{284} of the nascent peptide.[34]

The mature human uPAR consists of three repeats each apparently representing a different domain, with the NH_2-terminal amino acids (aa) 1-87 constituting the uPA binding domain.[28] These repeats exhibit low level internal homology and are characterized mainly by a unique pattern of cysteine residues and may also serve as three autonomous domains. These uPAR repeats share strong amino acid sequence homology with T cell activating antigen or Ly-antigens, squid brain glycoprotein and with MIRL/CD59.[37-39] A distinct feature common to all these proteins is the presence of a glycolipid membrane anchor. Furthermore, all members of this newly emerging superfamily of proteins have a single cysteine rich domain. Therefore these uPAR repeats are believed to have arisen from internal triplication of a common ancestral gene. Finally mapping of exact disulfide bridging of uPAR has further confirmed the close relationship of these proteins, since a similar pattern of disulfide bridging is also seen in the MIRL/CD59 protein.[40]

Recently two cDNAs encoding mouse uPAR have been characterized.[33] The first (mouse uPAR-1) is homologous to the human receptor and similarly to human uPAR, undergoes COOH terminal processing to become GPI-anchored at Gly_{275}. The second (mouse uPAR-2) encodes a 199 residue protein which shares the first 133 residues with uPAR-1.[33] Both contain a putative ligand binding domain but mouse uPAR-2 lacks the region involved in GPI-mediated membrane anchoring, suggesting that this may represent a secreted uPA binding protein. The two mRNAs species appear to arise by alternate splicing from a single gene. Differential expression of these alternately spliced variants of uPAR has been shown in the mouse gastrointestinal tract by in situ hybridization with riboprobes corresponding to the region distinct to the mouse uPAR-2. These studies have demonstrated the expression of both receptor subtypes in the epithelium of luminal and gastric mucosa in the mouse fundus and antrum. However muPAR-1 is specifically expressed over the luminal epithelium where no muPAR-2 expression is detected. More recently a human uPAR-2 has also been identified.[41] The human and mouse uPARs are species specific with respect to ligand binding which is due to critical differences in the aa (1-87) regions of the two receptors.

In order to understand the osteoblast selective growth factor-like effects of uPA we have isolated and characterized multiple isoforms of the rat uPAR from the non-transformed rat osteoblast cell line CFK-1. These studies have led to the isolation of three isoforms of rat uPAR. Rat uPAR-1 and 2 are homologous to the human and mouse uPAR-1 and 2, respectively. However from these rat osteoblasts we have isolated an additional uPAR isoform ruPAR-1b. This receptor isoform is identical to the ruPAR-1, with the exception of a critical single amino acid change from Cys to Ser at position 71 in ruPAR-1b due to a single base mutation. Although the cloned mouse and rat uPARs have 87% homology, they have less amino acid homology (57%) with their human counterpart in the critical uPA binding region (aa 1-87).[42] These interspecies amino acid differences may be responsible for the species-specificity of ligand binding observed by us and others.[18] The mRNAs for rat uPAR-1 and 2 encode mature proteins of 304 and 196 amino acids, respectively. Like the mouse, rat uPAR-2 arises from alternate splicing in its second internal repeat. This is in contrast to the human uPAR where this alternate splicing occurs in the third internal repeat.

The availability of this sequence information on various forms of human, rat and mouse uPAR should prove helpful in elucidating the basis of the species selectivity of uPA-binding and in evaluating the diverse roles of uPA in promoting cancer metastasis.

ACKNOWLEDGMENTS

This work was supported by the Medical Research Council of Canada (MRC) Grant MT-12609 and MT-10630. S.A. Rabbani is a recipient of scholarships from both the MRC of Canada and the Cancer Research Society of Canada.

REFERENCES

1. Reich R. Activation of plasminogen: a general mechanism for producing localized extracellular proteolysis. In: Berlin RD, Herman M, Lepow IH, Tanzer JM, eds. Molecular Basis for Biological Reproductive Processes. New York: Academic Press 1978; 155-69.
2. Belin D, Vassalli J-D, Combepine C et al. Cloning, nucleotide sequencing and expression of cDNAs encoding mouse urokinase-type plasminogen activator. Eur J Biochem 1985; 148:225-32.
3. Pennica D, Holmes WE, Kohr WJ et al. Cloning and expression of human tissue-type plasminogen activator cDNA in *E. coli*. Nature (Lond.) 1983; 301:214-21.
4. Stoppelli MP, Corti A, Soffientini A et al. Differentiation-enhanced binding of the amino-terminal fragment of human urokinase plasminogen activator to a specific receptor on U937 monocytes. Proc Natl Acad Sci USA 1985; 82:4939-43.
5. Mignatti P, Robbins E, Rifkin DB. Tumor invasion through the human amniotic membrane: requirement for a proteinase cascade. Cell 1986; 47:487-98.
6. Ossowski L. *In vivo* invasion of modified chorioallantoic membrane by tumor cells: the role of cell-surface bound urokinase. J Cell Biol 1988a; 107:2437-45.
7. Testa JE, Quigley JP. The role of plasminogen activator in aggressive tumor behavior. Cancer Metastasis Rev 1990; 9:353-67.
8. Fibbi G, Ziche M, Morbidelli L et al. Interaction of urokinase with specific receptors stimulates mobilization of bovine adrenal capillary endothelial cells. Exp Cell Res 1988; 179:385-95.
9. Sprengers ED, Kluft C. Plasminogen activator inhibitors. Blood 1987; 69:381-7.
10. Dano K, Andreasen PA, Grondahl-Hansen J et al. Plasminogen activators, tissue degradation, and cancer. Adv Cancer Res 1985; 44:139-266.
11. Laiho M, Keski-Oja J. Growth factors in the regulation of pericellular proteolysis: a review. Cancer Res 1989; 49:2533-53.

12. Wun T-C, Ossowski L, Reich E. A proenzyme form of human urokinase. J Biol Chem 1982; 257:7262-8.
13. Appella E, Robinson EA, Ullrich SJ et al. The receptor-binding sequence of urokinase: a biological function for the growth factor module of proteases. J Biol Chem 1987; 262:4437-40.
14. Verde P, Stoppelli MP, Galeffi P et al. Identification and primary sequence of an unspliced human urokinase poly(A)$^+$ RNA. Proc Natl Acad Sci USA 1984; 81:4727-31.
15. Au YPT, Wang TW, Clowes AW. Nucleotide and deduced amino acid sequences of baboon urokinase-type plasminogen activator. Nucleic Acids Research 1990; 18:3411.
16. Rabbani SA, Rajwans N, Murthy KK. PCR cloning of rat urinary plasminogen activator (uPA): implications for growth factor like activity in osteoblasts. J Bone Min Res 1993; 8:A76, S135.
17. Henderson BR, Tansey WP, Phillips SM et al. Transcriptional and posttranscriptional activation of urokinase plasminogen activator gene expression in metastatic tumor cells. Cancer Res 1992; 52:2489-96.
18. Rabbani SA, Desjardins J, Bell AW et al. An amino-terminal fragment of urokinase isolated from a prostate cancer cell line (PC-3) is mitogenic for osteoblast-like vells. Biochem Biophys Res Commun 1990; 173:1058-64.
19. Rabbani SA, Mazar AP, Bolivar I et al. Structural requirements for the growth factor activity of the amino-terminal domain of urokinase. J Biol Chem 1992; 267:14151-6.
20. Buko AM, Kentzer EJ, Petros A et al. Characterization of a post-translational fucosylation in the growth factor domain of urinary plasminogen activator. Proc Natl Acad Sci USA 1991; 88:3992-6.
21. Franco P, Mastronicola MR, De Cesare D et al. Separation and characterization of nonphosphorylated and serine-phosphorylated urokinase: catalytic properties and sensitivity to plasminogen activator inhibitor type 1. J Biol Chem 1992; 267:19369-72.
22. Henkin J, Dudlak D, Beebe DP et al. High sialic acid content slows prourokinase turnover in rabbits. Thromb Res 1991; 15:15-25.
23. Hollas W, Blasi F, Boyd D. Role of urokinase receptor in facilitating extracellular matrix invasion by cultured colon cancer. Cancer Res 1991; 51:3690-5.
24. Vassalli J-D, Baccino D, Belin D. A cellular binding site for the Mr 55,000 form of the human plasminogen activator, urokinase. J Cell Biol 1985; 100:86-92.
25. Stoppelli MP, Corti A, Soffientini A et al. Differentiation-enhanced binding of the amino-terminal fragment of human urokinase plasminogen activators to a specific receptor on U937 monocytes. Proc Natl Acad Sci USA 1985; 82:4939-43.
26. Appella E, Robinson EA, Ullrich SJ et al. The receptor-binding sequence of urokinase: a biological function of the growth-factor module of proteases. J Biol Chem 1987; 262:4437-40.

27. Blasi F. Urokinase and urokinase receptor: a paracrine/autocrine system regulating cell migration and invasiveness. Bio Essays 1993; 15:105-11.
28. Ploug M, Behrendt N, Lober D et al. Protein structure and membrane anchorage of the cellular receptor for urokinase-type plasminogen activator. Sem Thromb Hemost 1991; 17:183-93.
29. Kirchheimer JC, Wojta J, Christ G et al. Proliferation of a human epidermal tumor cell line stimulated by urokinase. FASEB J 1987; 1:125-8.
30. Estreicher A, Mühlhauser J, Carpentier J-L et al. The receptor for urokinase type plasminogen activator polarizes expression of the proteases to the leading edge of migrating monocytes and promotes degradation of enzyme inhibitor complexes. J Cell Biol 1990; 111:783-92.
31. Cubellis MV, Wun T-C, Blasi F. Receptor-mediated internalization and degradation of urokinase is caused by its specific inhibitor PAI-1. EMBO J 1990; 9:1079-1085.
32. Roldan AL, Cubellis MV, Masucci MT et al. Cloning and expression of the receptor for human urokinase plasminogen activator, a central molecule in cell surface, plasmin dependent proteolysis. EMBOJ 1990; 9:467-74.
33. Kristensen P, Eriksen J, Blasi F et al. Two alternatively spliced mouse urokinase receptor mRNAs with different histological localization in the gastrointestinal tract. J Cell Biol 1991; 115:1763-71.
34. Borglum AD, Byskov A, Ragno P et al. Assignment of the urokinase-type plasminogen activator receptor gene (PLAUR) to chromosome 19q 13.1-q13.2. Am J Hum Genet 1992; 50:492-7.
35. Ploug M, Rønne E, Behrendt N et al. Cellular receptor for urokinase plasminogen activator: carboxyl-terminal processing and membrane anchoring by glycosyl-phosphatidylinositol. J Biol Chem 1991; 226:1926-33.
36. Thomas JR, Dwek RA, Rademacher TW. Structure, biosynthesis and function of glycosylphosphatidylinositols. Biochemistry 1990; 29:5413-5422.
37. Reiser H, Coligan J, Palmer E et al. Cloning and expression of a cDNA for the T-cell-activating protein TAP. Proc Natl Acad Sci USA 1988; 85:2255-9.
38. Williams AF, Tse AGD, Gagnon J. Squid glycoproteins with structural similarities to Thy-1 and Ly-6 antigens. Immunogenetics 1988; 27:265-72.
39. Davies A, Simmons DL, Hale G et al. CD59, an Ly-6-like protein expressed in human lymphoid cells, regulates the action of the complement membrane attack complex on homologous cells. J Exp Med 1989; 170:637-54.
40. Ploug M, Kjalke M, Rønne E et al. Localization of the disulfide bonds in the NH_2-terminal domain of the cellular receptor for human urokinase-type plasminogen activator: A domain structure

belonging to a novel superfamily of glycolipid-anchored membrane proteins. J Biol Chem 1993; 268:17539-46.
41. Pyke C, Eriksen J, Solberg H et al. An alternatively spliced variant of mRNA for the human receptor for urokinase plasminogen activator. FEBS 1993; 326:69-74.
42. Rabbani SA, Rajwans N, Achbarou A et al. Isolation and characterization of multiple isoforms of the rat urokinase receptor in osteoblasts. FEBS Lett 1994; 338:69-74.

CHAPTER 9

The Metalloproteinases and Their Inhibitors

Margaret Durko and Pnina Brodt

INTRODUCTION

The extracellular matrix (ECM) constitutes a three-dimensional multi-functional protein meshwork which acts as a scaffolding to support and maintain tissue structure, as a barrier separating the different tissue compartments from each other and as a source of molecular signals regulating cell polarity, differentiation and growth. The two types of extracellular matrix structures which are of major interest to this review are the basement membrane-type ECM and the interstitial stroma. These differ in their molecular composition, location within the tissue and function. The basement membrane, which forms the interface between the organ parenchyma and the interstitial stroma and ensheathes blood vessels, muscle cells and nerves, is composed mainly of types IV and V collagen, fibronectin, laminin, entactin, and heparan sulfate proteoglycans.[1] Adjacent to the basement membrane (and facing the lamina densa) is the interstitial stroma which is associated with connective tissue cells and is composed mainly of fibrous components such as types I, III and V collagen, elastin, proteoglycans and glycoproteins such as fibronectin.[2] Under normal physiological conditions, these matrices form dense barriers impenetrable to most cells. However, under specific but diverse conditions the matrix can become locally permeable to cell movement. These include on one hand normal physiologic processes such as

Cell Adhesion and Invasion in Cancer Metastasis, edited by Pnina Brodt.

embryogenesis, inflammation, wound healing and tissue remodeling and on the other, pathological processes such as tumor invasion and metastasis.[3,4]

In the past three decades the biochemical mechanisms involved in matrix dissolution have been the focus of intense investigation leading to identification of several families of proteinases with matrix-degrading activities. One family of enzymes whose members play a major role in cell migration and tissue invasion in normal and pathological processes are the metalloproteinases. Under normal physiological conditions, the activity of these enzymes is tightly controlled at the level of gene expression, proenzyme activation and interaction with specific inhibitors such as the tissue inhibitors of metalloproteinases (TIMPs).[5,6] In malignancy, however, this tight regulation of proteinase expression and activity appears to be disrupted.[7]

This chapter provides a brief review of the common properties of the metalloproteinases and a discussion of the individual members of the family. For more in depth analysis the reader is referred to several excellent reviews published elsewhere.[4,8-10]

COMMON STRUCTURAL MOTIFS OF THE METALLOPROTEINASES

The basement membrane and interstitial stroma are rich in collagens. In their native forms the collagens can only be degraded by specific collagenases. These collagenases, together with a second group of enzymes—the stromelysins, constitute the family known as matrix metalloproteinases (MMPs). Members of this family are zinc-dependent endopeptidases with a broad spectrum of proteolytic activity against several components of the extracellular matrix. Included in the family are: interstitial collagenase (also known as type I collagenase and MMP-1), neutrophil collagenase (MMP-8), a 72 kDa type IV collagenase (gelatinase A or MMP-2), a 92 kDa type IV collagenase (gelatinase B or MMP-9), stromelysin 1 (MMP-3, transin, proteoglycanase), stromelysin 2 (MMP-10, transin-2), stromelysin 3 (MMP-11), and matrilysin (PUMP-1, MMP-7).[8,9] Three newly identified members of the family which have not yet been fully characterized are human macrophage metalloelastase (HME, MMP-12),[11] a membrane-type matrix metalloproteinase (MT-MMPm, MMP-14) and collagenase-3 (MMP-13)[12,13] (see Table 9.1 for summary). These enzymes share

Table 9.1. The metalloproteinase family

Group	Enzyme	Molecular Mass (kDa)	Substrates	References
I	**Type I collagenase** Interstitial collagenase MMP-1	57 (latent) 52 (latent) 47 (active) 42 (active)	Collagen types I, II, III	32, 40
	Neutrophil collagenase MMP-8	75 (latent) 57 (active)	Collagen types I, II, III	52
	Collagenase 3 MMP-13	66 (latent) 55 (active)	Collagen type I	13
II	**72 kDa type IV collagenase** Gelatinase A MMP-2	72 (latent) 66 (active)	Collagen types IV, V, VII, X, gelatins, elastin, fibronectin	29
	92 kDa type IV collagenase Gelatinase B MMP-9	92 (latent) 84 (active)	Collagen types IV, V, VII, X, gelatins, elastin, fibronectin, laminin	93
III	**Stromelysin-1** MMP-3 45 (active)	60 (latent) 57 (latent) 28 (active)	Collagen types IV, V, IX, proteoglycans, laminin, fibronectin, casein, elastin	 54
	Stromelysin-2 MMP-10	53 (latent) 47 (active) 28 (active)	Collagen types III, IV, V, fibronectin gelatins	66, 67
	Stromelysin-3		unknown substrate possibly other?	68a, 68b
	Matrilysin MMP-11 PUMP-1 MMP-7	28 (latent) 19 (active)	Casein, gelatins, fibronectin, laminin, elastin, proteoglycans	44, 67, 73
IV	**Metalloelastase** HME MMP-12	54 (latent) 45 (active) 22 (active)	Elastin	11
V	**MT-MMP** (Membrane-type matrix metalloproteinase) MMP-X1 MMP-14	66 (deduced)	Pro-gelatinase A activator	12

a similar amino-terminal region consisting of a signal peptide and a propeptide domain commonly cleaved on activation of the proenzyme. In addition they share a common zinc-binding domain and a hemopexin-like domain (which has a sequence homology to hemopexin, a heme-binding serum protein) in the carboxyl-terminus (see Fig. 9.1). The hemopexin-like domain has four short tandem repeats which also occur twice in the ECM protein vitronectin. This domain may be responsible for matrix binding and is thought to play a role in the substrate specificity of the enzymes[9,14,15] as its deletion in interstitial collagenase resulted in the loss of enzyme specificity for native, type I collagen.[14] Both MMP-2 and MMP-9 have an additional, 175 amino acids (a.a.) fibronectin-like, collagen-binding domain consisting of three head-to-tail repeats of 58 amino acids each which are homologous to the type II motif of fibronectin.[16] This domain is located amino terminally and adjacent to the zinc-binding sequence.[8] MMP-9 also has a unique α2 (V) collagen-like domain which is inserted between the catalytic and hemopexin-like domains[8,9] (see schematic representation in Fig. 9.1).

Whereas in normal cells the repertoire of metalloproteinases is restricted, malignant cells of different origins such as breast, lung, prostate and salivary-gland adenocarcinomas, fibrosarcomas,

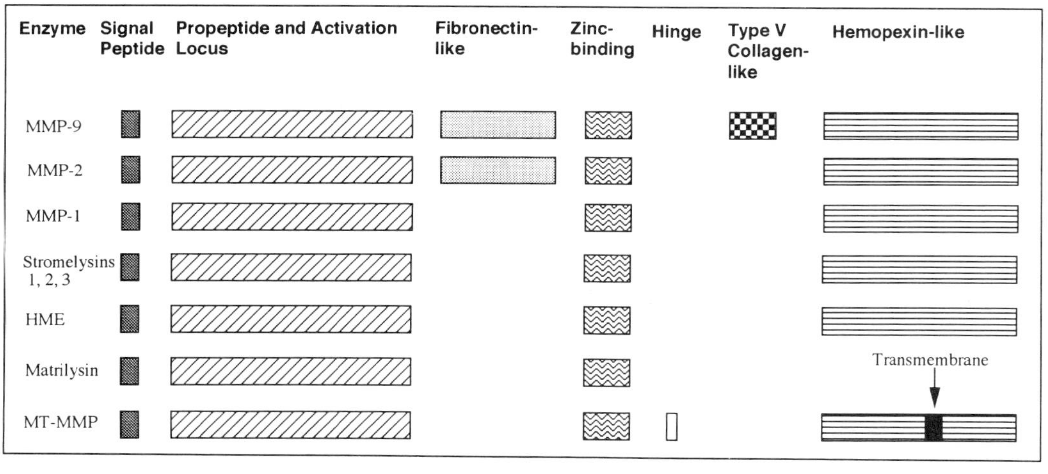

Fig. 9.1. Structural and functional domains of the metalloproteinases. Shown is a schematic representation of common and unique domains in the metalloproteinases. In the N-terminal region, the common domains include a signal peptide (cleaved upon secretion), a propeptide (cleaved upon activation), and an activation locus which together with the zinc-binding domain is required for proteolytic activity. The hemopexin-like domain in the C-terminus is found in all the metalloproteinases except matrilysin.

neuroblastomas and melanomas have been found to co-express several members of the MMP gene family simultaneously.[17-21]

MECHANISMS OF METALLOPROTEINASE ACTIVATION

Metalloproteinases are secreted as latent proenzymes (zymogens) and are normally activated in the extracellular milieu. The precise mode of enzyme activation in vivo has been the subject of some controversy and is thought to be mediated either through autoproteolysis or by the proteolytic action of other enzymes.[5,12,22,23] In vitro, the enzymes can be activated by a variety of reagents. These include proteases (e.g. trypsin, plasmin) and conformational perturbants such as SDS, heavy metals, oxidants, disulfide

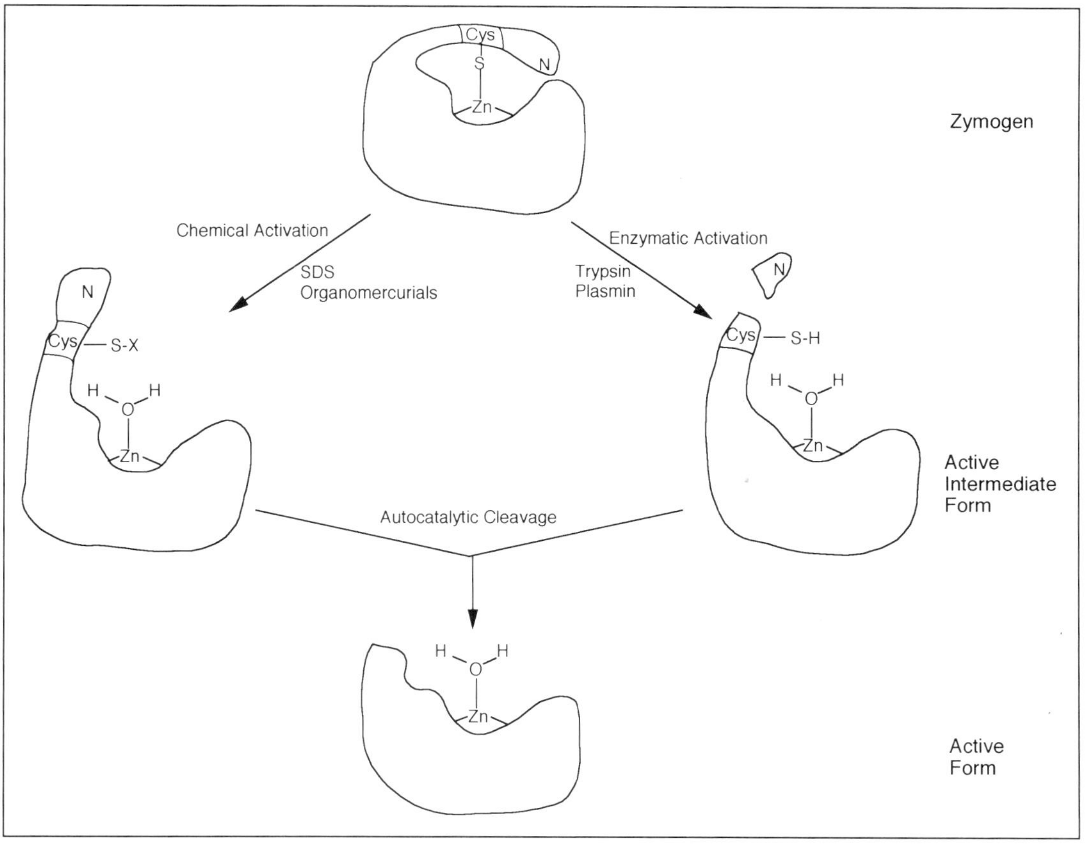

Fig. 9.2. The stepwise processing of the metalloproteinases. The active intermediate forms are generated through a conformational change in the molecule (chemical activation) or by proteolytic cleavage of the propeptide at a site located N-terminally to the Cys residue (enzymatic activation). The final active form is a product of autocatalytic cleavage at a site located C-terminally to the Cys residue.

compounds and sulfhydryl-alkylating agents—all of which convert the proenzyme to an active intermediate enzyme form.[8,9,16] The intermediates are then autolytically cleaved to remove the 80 residue propeptide fragment from the N-terminus of the proenzyme, generating a stable active enzyme[9,16,24] (see diagram in Fig. 9.2). A well-conserved amino acid sequence PRCGVPD with a cysteine residue bound to the zinc ion is located immediately upstream of the autolytic cleavage site. This configuration is thought to be responsible for maintaining enzyme latency.[16] During activation, as the propeptide containing an unpaired cysteine residue within the conserved motif PRCGVPD is cleaved from the proenzyme, the cysteine-zinc bond is disrupted.[9,25] This dissociation of the cysteine residue from the zinc atom and its replacement by water is believed to expose the active site and convert the enzyme to a catalytically active form, a process termed the "cysteine switch" mechanism[16,25] (see Fig. 9.2).

MEMBERS OF THE METALLOPROTEINASE FAMILY

Interstitial (Type I) Collagenase (MMP-1) and Related Molecules

MMP-1

Function

MMP-1 initiates the breakdown of types I, II and III fibrillar collagens which together constitute the most abundant proteins in the body.[9,26] It therefore plays a critical role in tissue remodeling which occurs under both physiological and pathological conditions. Cleavage of all three collagen α chains takes place at a single locus by hydrolysis of the peptide bond between residues 771 and 772 which is located immediately following a Gly residue. This produces two proteolytic fragments approximately 3/4 and 1/4 the size of the original molecule.[27] In addition to collagens, MMP-1 also has a limited capacity to degrade proteoglycans.[28] Fibroblast MMP-1 has also been shown to have some degradative activity on type IV collagen but the cleavage products are different from those of the type IV collagenase (MMP-2).[29] In addition it has recently been shown that the enzyme can promote the activation of MMP-2.[30]

Expression

Full length cDNA for human synovial and dermal cell collagenases have been cloned and their sequences were found to be identical and code for a 2.5 kb mRNA transcript.[31,32] A cDNA isolated from a melanoma cDNA library was found to be identical to the normal counterpart with the exception of two base pair substitutions.[33] The nucleotide sequence for MMP-1 is well-conserved through the species. For example the porcine collagenase cDNA sequence has a high homology to the human and rabbit enzymes (84% and 83.4 % respectively)[34] which in turn have 86% homology with each other.[35] The gene encoding human fibroblast interstitial collagenase is located on chromosome 11, spans 8.2 kb and consists of 10 exons and 9 introns,[36,37] whereas the rabbit synovial cell collagenase gene spans 9.1 kb consisting also of 10 exons and 9 introns.[35] The enzyme can be secreted or associated with the plasma membrane in both neoplastic and normal cells.[38,39] The enzyme produced by human dermal fibroblasts is synthesized as a 54,092 Da pre-proenzyme with a 19 amino acid signal peptide.[32] These cells secrete two proenzyme species of 52 and 57 kDa,[32,40] the larger molecule resulting from post-transcriptional modification of the major 52 kDa molecule through the addition of N-linked complex oligosaccharides. Similar patterns of expression and secretion have also been reported for endothelial cells and fibroblasts derived from different organs such as the colon, cornea, gingiva, lung, and fetal skin.[40]

Activation

Both secreted forms can be activated by limited digestion with trypsin or plasmin generating their respective 42 and 47 kDa active enzymes.[40,41] Enzymatic digestion of the rabbit fibroblast collagenase with trypsin yields similar products.[42] On the other hand, activation by organomercurials causes a conformational change leading to collagenolytic activity without concomitant loss of molecular mass. This however is followed by auto-proteolytic cleavage and the appearance of two lower molecular weight species of 44 and 42 kDa.[43] Plasmin or trypsin-activated collagenase can be further processed to a fully active 40 kDa enzyme by activated stromelysin[41]—an example of metalloproteinases engaged in a proteolytic cascade. This activation, which involves the removal of 15 amino acid residues from the carboxyl end of the enzyme results in a 5- to 8-fold increase in the specific activity of the enzyme.[41] Activated gelatinase A can also increase the activity of MMP-1 by cleaving the peptide bond between amino acids 80 and 81.[30] In addition, matrilysin has also been reported to activate human type I procollagenase.[44]

Regulation

The synthesis of collagenase can be induced by a variety of mediators including cytokines such as IL-1 and TNFα, growth factors such as EGF, PDGF and bFGF and chemical modulators such as phorbol esters, cytochalasin B, and calcium phosphate crystals.[34,40,45,46] Both cytokines (IL-1) and growth factors (EGF, bFGF) can induce expression of collagenase and stromelysin simultaneously.[47,48] On the other hand, biosynthesis can be inhibited by corticosteroids, retinoids and TGFβ.[33,48,49] TGFβ has an inhibitory effect on the basal expression and can also inhibit induction of collagenase mRNA by cytokines and growth factors.[34]

Neutrophil Collagenase (MMP-8)

Function

MMP-8 is a distinct protease which has essentially the same substrate specificity as MMP-1 and cleaves types I, II and III collagens, with a preference for type I collagen.[50] Similarly to MMP-1, it initiates degradation of native interstitial collagen by cleavage at a single site in the collagen triple helix producing two fragments which are 3/4 and 1/4 the size of the intact molecule. The region responsible for the substrate specificity of the enzyme has been localized to the carboxyl-terminal domain.[51]

Expression and activation

A full length MMP-8 cDNA encoding a 3.3 kb mRNA transcript with a predicted 467 residue gene product was isolated from a human peripheral leukocytes cDNA library.[52] During neutrophil maturation in the bone marrow, this enzyme is stored in intracellular granules as a latent, 75 kDa highly glycosylated molecule. The proenzyme is converted to the 57 kDa activated form by proteolytic processing.[52]

Amino acid sequence analysis revealed that this enzyme has 57 and 52% homologies respectively to human fibroblast MMP-1 and stromelysin 1.[52]

Collagenase-3 (MMP-13)

Function

The information available on this newly described enzyme is still limited but type I collagen has been identified as the major substrate.[13]

Expression and activation

The cDNA cloned from human breast carcinoma specimens has a 1413bp open reading frame and codes for a 471 aa protein with a predicted molecular mass of 53,759 Da.[13] Sequence analysis revealed that this protein shares a 50.5% homology with human neutrophil collagenase and a 36% homology with human stromelysin-3. However, the most extensive homology (86%) was found with rat and mouse type I collagenases.[13] A 65 kDa species (possibly a glycosylated form of the polypeptide) was detected in medium conditioned by cells which were infected with the recombinant vaccinia virus expressing the complete cDNA sequence. Upon activation with organomercurials, a lower molecular weight species of 55 kDa appeared.[13] RT-PCR analysis could detect collagenase-3 mRNA in breast carcinoma specimens but not in breast cancer lines or in a variety of normal tissues analyzed including normal breast epithelium, mammary fibroadenomas, liver, placenta, ovary, uterus, prostate and parotid gland[13] suggesting that the expression of this enzyme is associated with malignant transformation in vivo. In breast carcinoma tissue, three mRNA species of 2, 2.5 and 3 kb have been observed. These transcripts are due to utilization of three different polyadenylation sites present in the 3'-flanking region of collagenase-3.[13]

The Stromelysins

Three stromelysins, which were designated stromelysins 1, 2 and 3, have been identified to date. Matrilysin has also been included in this section because its substrate specificity is similar to that of the stromelysins.[44,53]

Stromelysin-1 (MMP-3)

Function

Stromelysin-1 can degrade a broad range of substrates including fibronectin, laminin, types IV, V, and IX collagens (cleaved in the non-helical region), the protein core of proteoglycans, casein, gelatin and elastin.[28,54,55] In addition, activated stromelysin is a potent activator of type I collagenase increasing its activity by up to 12-fold relative to trypsin-mediated activation.[55-57]

Expression

Enzyme expression has been observed in different cells including human dermal fibroblasts, fetal skin and lung fibroblasts, synovial cells and keratinocytes.[54,58,59] The gene for human stromelysin is located on chromosome 11 and spans 14 kb.[37,58] A 2.3 kb mRNA transcript has been detected in human dermal and rabbit synovial fibroblasts.[54,60] The complete primary protein structure has been elucidated. It consists of 477 aa residues[54,59] with a predicted molecular mass of 53,977 Da.[54] Human stromelysin shares a 55% homology with human type I collagenase[59,61] but is more closely related to rat transin (75% homology).[54,59,62]

Activation

Human prostromelysin is secreted as two proenzymes of 57 and 60 kDa each. The minor 60 kDa form is the result of partial N-linked glycosylation.[54] In vitro, the major 57 kDa proenzyme can be converted into a 45 kDa active enzyme by trypsin,[54] plasmin and organomercurials[55,56] or by heating to 55° C. Two active forms of 45 and 28 kDa were purified from medium of human rheumatoid synovial cells.[63] The activation pathway in vivo is still, however, unknown.[55]

Regulation

The regulation of stromelysin expression and activity appears to be cell-type specific. Thus cultured human fibroblasts derived from skin, gingiva, colon and cornea secrete stromelysin constitutively and the expression can be upregulated by phorbol esters.[54] Human keratinocytes on the other hand express low constitutive levels of stromelysin and the expression cannot be stimulated by phorbol esters.[54] The expression of stromelysin and type I collagenase is often, but not always, coordinately regulated. In some cells such as rabbit synovial cells, alveolar macrophages and brain capillary endothelial cells, both enzymes could be coordinately upregulated by phorbol esters and cytochalasin B.[60] Similarly, expression of both enzymes in human foreskin and synovial fibroblasts could be coordinately induced by phorbol esters, EGF and IL-1β[58,59,64] while in human and rabbit synovial fibroblasts transcription of both enzymes could be downregulated with retinoic acid and glucocorticoids.[59,65] In other cells, however, the expression of stromelysin does not parallel that of collagenase. For example, in melanoma, fibrosarcoma and SV-40 transformed human fetal lung fibroblasts, upregulated expression of the collagenase but not of stromelysin-1 was observed[54] and the latter could not be stimulated by phorbol esters. Moreover, IFNγ downregulates the expression of stromelysin and consequently reduces the levels of activated procollagenase but it does not affect procollagenase protein expression *per se.*[5]

Stromelysin-2 (MMP-10)

Function

MMP-10 can degrade fibronectin, casein, and type I, III, IV and V gelatins but has only weak activity against types III, IV and V native collagens.[9,66] Similarly to stromelysin-1, it can also activate type I procollagenase in vitro.[66]

Expression and activation

The gene for human stromelysin-2 spans 14 kb[58] and encodes a 1.9 kb mRNA transcript[67] with a predicted 476 aa gene product.[67] The actual molecular mass of the secreted protein is 53 kDa[9] and this gives rise to two molecules of 47 and 28 kDa each, upon activation with agents such as organomercurials.[9,66] The stromelysin-2 protein shares a 52% homology with type I collagenase[67] and a 78% overall homology with stromelysin-1.[58] However, the expression of stromelysins 1 and 2 appears to be regulated by distinct mechanisms.[58]

Regulation

Distinct regulation and differential expression of the stromelysins have been demonstrated in rheumatoid synovial cells, human foreskin fibroblasts, rabbit synovial fibroblasts[58] and some tumor cells.[67] In human rheumatoid synovial and foreskin fibroblasts, the expression of stromelysin-1 could be upregulated by phorbol ester, EGF or IL-1β whereas expression of stromelysin-2 was either minimally increased (phorbol ester) or unaffected (EGF, IL-1β).[58]

Stromelysin-3 (ST3, MMP-11)

Expression and regulation

A full-length cDNA for ST3 was first isolated from a cDNA library constructed from a breast carcinoma biopsy containing both neoplastic and stromal components.[68] The nucleotide sequence determined predicts a 488-residue protein. This putative protein has a relatively low sequence homology to stromelysins-1 (40%) and 2 (38%) or to type I collagenase (36%). Subsequently, enzyme production was shown to be restricted to stromal cells adjacent to the malignant tissue as seen in breast carcinomas and squamous cell carcinomas of the head and neck,[68,69] while no expression could be detected in benign lesions or in established breast cancer cell lines.[68] This has led to suggestions that this enzyme may be an important component of the reactive stroma, where it is often found in association with breast carcinoma tissue and may play a role in disease progression by facilitating tumor cell invasion.[68] Low levels of ST3 mRNA were detected in carcinomas of the colon, ovary, kidney and lung as well as during normal embryonic development but no message was found in several normal human adult tissues. ST3 gene expression in fetal fibroblasts could be modulated by phorbol esters as well as by several growth factors including PDGF, EGF and bFGF.[68] As expression of these growth factors is often upregulated in malignant cells,[70,71] it is possible that stromelysin-3 is a product of epithelial-stromal interactions occurring as a by-product of carcinoma progression. The substrate specificity of this enzyme has not to date been conclusively defined.

Matrilysin (Pump-1, MMP-7)

Function

Matrilysin is a broad spectrum proteinase which can degrade casein, gelatins (types I, III, IV, V), fibronectin, laminin, and elastin and activate type I collagenase.[44,53] In addition it has been shown to regulate urokinase plasminogen activator function by catalyzing the hydrolysis of the prourokinase at the Glu_{143}-Leu_{144} bond.[72]

Expression and activation

A matrilysin cDNA clone has been isolated from a cDNA library constructed from a pool of mRNAs derived from several human tumors and probed with a rat transin cDNA.[67] Human matrilysin gene spans 9.65 kb and is composed of six exons and 5 introns.[73] It codes for a 1. 2 kb mRNA transcript[44,74] and a protein of 267 a.a. residues with a predicted molecular mass of 27,938 Da.[44,67] This relatively small size is due to the absence of the C-terminal hemopexin-like domain in this enzyme.[9,67] A comparison of the sequence with corresponding regions of stromelysin-1 and interstitial collagenase revealed homologies of 49% and 44% respectively.[67] The 28 kDa latent form of the secreted protein is prone to activation with organomercurials and following autocleavage gives rise to two species of 21 and 19 kDa each.[44,53,72] The enzyme is expressed in mononuclear phagocyte precursors but not in terminally differentiated phagocytes such as alveolar macrophages,[53] and may play a role in monocyte homing into specific tissues. Matrilysin is expressed in human malignancies such as prostate, gastric and colon carcinomas and in colon carcinoma-derived cell lines.[73-75] It was shown to play a role in the invasion of prostate cancer.[74]

Regulation

Phorbol ester and EGF positively regulate matrilysin as was shown in human colon carcinoma cell lines in which the enzyme was constitutively expressed.[73] The constitutive expression of matrilysin in immature bone marrow-derived promonocytes or peripheral blood monocytes could be markedly increased by stimulation with phorbol esters and mitogens such as lipopolysaccharides and Con A.[53]

Type IV Collagenases (Gelatinases)

The 72 kDa type IV collagenase (gelatinase A, MMP-2)

Function

The role of MMP-2 in malignancy has received much attention in recent years.[8] The major substrate for this enzyme is type IV collagen—the major basement membrane protein. In addition, it has also been implicated in the degradation of types V, VII and X collagens, gelatins, elastin and fibronectin.[9,29,76,77] These substrates bind to the enzyme via the fibronectin-like, collagen-binding domain.[29] There is presently some controversy regarding the ability of MMP-2 to degrade native, full length type IV collagen. Thus while the enzyme has been shown to initiate the degradation of type IV collagen in a pepsin-resistant portion of the molecule,[29] other evidence suggests that it can digest only precleaved or conformationally altered forms of collagen and may require the cooperation of other enzymes to degrade the native form.[18]

Expression

The MMP-2 gene is located in the q21 region of chromosome 16. It spans 27 kb and consists of 13 exons and 12 introns.[9,78,79] The mRNA transcript is 3.1 kb long and encodes a secreted, 72 kDa proenzyme.[29] The sequence of human type IV procollagenase shows limited homology with human interstitial collagenase and stromelysin-1.[24,29] Expression of this enzyme while detectable in normal cells such as fibroblasts,[29,80,81] is significantly upregulated in many tumor types[17,19,82-84] suggesting that it is involved in the progression of malignant disease.[83,84] Indeed enhanced enzyme production has been linked to viral transformation as seen in SV-40-transformed human lung fibroblasts[29] and to expression of several oncogenes such as *src* and *ras*[29,85] in a variety of transformed cells including *H-ras* transformed human bronchial epithelial cells and rat embryo fibroblasts[29,86] or Rous sarcoma virus-transformed chicken embryo fibroblasts.[85]

Activation

Activation of the 72 kDa proenzyme by agents such as organomercurials results in the autocatalytic cleavage of an amino-terminal domain giving rise to two activated forms of 66 and 62 kDa each.[24,29,85] Activated forms of the enzyme can also be generated by other MMPs such as matrilysin[23] and membrane type MMP (MT-MMP).[12] Interstitial collagenase can also promote the activation by generating a 68 kDa intermediate which is in turn converted to the 66 kDa active form by autolytic activation.[30] Unlike other metalloproteinases, trypsin was found in some studies to be a poor activator of the MMP-2 proenzyme.[29,85] However, the conversion of the 72 kDa proenzyme to a 62 kDa gelatinase by a tumor associated trypsin (TAT-2) has recently been described.[85a] There is some evidence to suggest that the enzyme can be active in a membrane associated form as it was isolated in the plasma membrane fraction of some tumor cells[87] and two human breast adenocarcinoma cell lines were able to bind exogenous enzyme.[88] The membrane-dependent activation results in N-terminal cleavage of the proenzyme generating a 64 kDa intermediate which is subsequently converted to the 62 kDa active form.[89] TIMP-2 (see below) blocks this activation by forming a complex with the 72 kDa proenzyme while TIMP-1 forms a complex with the intermediate form and prevents its autocatalytic conversion.[89]

Regulation

The transcription of MMP-2 appears to be regulated differently than that of stromelysin and interstitial collagenase as can be predicted from the differences in the promoter region[78,82] (see below). Phorbol ester which stimulates MMP-1 and stromelysin expression was shown to have the reverse effect on the expression of MMP-2 while TGFβ1 stimulates the latter but inhibits the former.[90,91] Of particular relevance to the focus of this book are the findings that expression and secretion of MMP-2 could be positively regulated by ligation of the vitronectin receptor $\alpha_v\beta_3$. This was demonstrated by using immobilized ligand or monoclonal antibodies to cross-link the integrin receptors.[80,92]

The 92 kDa type IV collagenase (gelatinase B, MMP-9)

Function

MMP-9 is similar to MMP-2 in both structure and substrate specificity,[9,93] but is differentially regulated.[91] Among the known substrates for this enzyme are gelatin, elastin, fibronectin, proteoglycans, laminin and types IV, V, VII and X collagens.[9,77,91,93,94] As is the case for MMP-2, it remains unclear whether MMP-9 can initiate degradation of the native full length type IV collagen[18] or can act only on a conformationally modified molecule.

Expression

The complete structure of the human MMP-9 gene has been elucidated. It is located on human chromosome 16, spans 26 kb and consists of 13 exons and 12 introns.[79,91] The mRNA transcript is 2.8 kb long.[93] The molecular mass of the preproenzyme is 78,426 Da but the enzyme is secreted as a 92 kDa molecule due to post-translational glycosylation.[93] Similarly to MMP-2, this enzyme contains a fibronectin-like collagen-binding domain[9,29] and in addition, a unique proline-rich collagen-like domain which has not been found in other members of the metalloproteinase family.[93] The enzyme is expressed by a variety of normal cells including human macrophages, epidermal keratinocytes, polymorphonuclear leukocytes and placental cytotrophoblasts.[93,95] It is not normally detectable in stromal cells such as lung or dermal fibroblasts[96] but its expression was found to be upregulated in chondrocytes of osteoarthritic cartilage, in SV-40 transformed human lung fibroblasts, in transformed rat embryo cells and in tumor cell lines suggesting that the enzyme plays a role in inflammation of the joints and in tumor progression.[18,93,94,97] Elevated plasma levels of this enzyme were also reported during pregnancy and in cancer patients.[98]

Activation

Activation in vitro by organomercurials, plasmin or stromelysin results in the removal of a 73 a.a. fragment from the N-terminus of the proenzyme and its conversion into an 84 kDa active species.[93,99,100] The enzyme associates preferentially with its inhibitor TIMP-1 through its carboxyl-terminal hemopexin-like domain.[99] This association can be prevented by the removal of as few as six amino acids from the carboxyl-terminus. The secreted procollagenase can form a covalent homodimer as well as a heterodimer with interstitial collagenase[99] and can also be found as a trimolecular complex with TIMP-1 and activated interstitial collagenase. The complex formed between the 92 kDa and interstitial collagenases retains enzymatic activity towards substrates of both enzymes (i.e. gelatin and fibrillar type I collagen, respectively) upon its activation with either plasmin or stromelysin.[99]

Regulation

Transformed or malignant cells and rheumatoid synovial fibroblasts frequently express the enzyme[96,99,100] and it can be stimulated in these cells by inducers such as phorbol esters, IL-1β, EGF and TNFα.[91,93,94,96,99-101] Normal dermal and lung fibroblasts do not produce MMP-9.[96] However, in dermal fibroblasts, expression could be induced by cytokines such as IL-1β, TNFα and lymphotoxin while growth factors such as PDGF and bFGF were ineffective. IL-1 also induces expression of MMP-9 in normal human cartilage explants and isolated chondrocytes.[94]

Human Macrophage Metalloelastase (HME, MMP-12)

Function and Expression

This proteinase is expressed in alveolar macrophages and has been shown to degrade elastin.[11] A full length cDNA for HME was cloned from a human alveolar macrophage cDNA library and contains an open reading frame of 1410 bp.[11] It encodes a 1.8 kb mRNA and a 54 kDa proenzyme. Two active forms are generated from the secreted proenzyme—a 45 kDa (after N-terminal processing) and a 22 kDa (after an additional C-terminal processing) molecule.[11] HME is 74% identical to murine macrophage metalloelastase (MME) at the nucleotide level and 64% identical at the amino acid level. HME has 49% a.a. identity to stromelysin-1 and interstitial collagenase.[11]

Membrane-Type Matrix Metalloproteinase (MT-MMP, MMP-X1, MMP-14)

Function and Expression

A full length cDNA for MT-MMP was cloned from a human placental cDNA library. It encodes a 4.5 kb mRNA transcript and a 582 a.a. protein with a molecular mass of 66 kDa.[12] The protein is closely related to other members of the MMP family but has an additional, putative transmembrane domain at the C-terminus which consists of 24 hydrophobic amino acids which are inserted into the hemopexin-like domain.[12] MT-MMP is expressed on the cell surface where it functions as a specific activator of progelatinase A.[12] mRNA transcripts for this enzyme were demonstrated in normal and malignant lung tissue but its role in malignancy remains to be confirmed.[12]

THE TISSUE INHIBITORS OF METALLOPROTEINASES (TIMPs)

Three members of this emerging family of inhibitors have been identified to date and were designated TIMP-1, 2 and 3. Together they provide a tightly regulated mechanism for control of MMP activation and function.

TIMP-1

Function

Among the metalloproteinases which are inhibited by TIMP-1 are the collagenases, MMP-3 and the gelatinases[99,102,103] which interact with the inhibitor in a stoichiometric (1:1 molar basis) manner.[15,103] Two functional domains have been identified in both TIMP-1 and TIMP-2;[104] the amino-terminal, inhibitory domain which is generic and interacts with all open MMP active sites and the carboxyl-terminal domain which is gelatinase specific. MMP-1 and stromelysin-1 require prior activation in order to bind TIMP-1[103] while MMP-9 can form a complex with the inhibitor in its latent, procollagenase form.[99,103] The latter association prevents the proteolytic activation of the proenzyme by stromelysin or plasmin[99] but not the autoactivation initiated by organomercurials.[103]

Expression

Human TIMP-1, a 28.5 kDa ubiquitous glycoprotein is produced by bone marrow derived cells such as platelets and macrophages[105,106] and by connective tissue cells and is found in human body fluids.[102,107] It has been detected in conditioned media derived from different cell cultures[19,47,48,108] but its production is markedly reduced in cultures of neoplastic cells.[19] The TIMP-1 gene has been mapped to the X chromosome.[109] The full length cDNA codes for a 0.9 kb transcript[110] and a 207 aa protein in which the first 23 residues form a signal peptide that is cleaved upon secretion.[107,111] The secreted form of the inhibitor consists of 184 a.a. residues, contains two sites of N-linked glycosylation and 12 cysteine residues which have been assigned to six disulfide bonds.[15,107] The non-glycosylated, recombinant TIMP-1 has a molecular mass of 20,711 Da and retains its inhibitory potential suggesting that the carbohydrates are not required for function.[107,112]

Regulation

The physiological regulators of TIMP-1 are cytokines such as IL-1 and growth factors such as EGF, bFGF and TGFβ. These, as well as the tumor promoter, phorbol ester,[47,48] have a stimulatory effect and substantially increase TIMP-1 secretion. The stimulatory activity of TGFβ requires cooperation with growth factors such as EGF and bFGF to super-induce inhibitor expression.[48] At the functional level the inhibitory activity of the protein is regulated by various serine proteases, including human neutrophil elastase, trypsin and α-chymotrypsin which may all mediate its destruction.[113]

In addition to its major known function as an inhibitor of metalloproteinase activity (or possibly as a consequence of this function), TIMP-1 has also been implicated in the regulation of other cellular functions.[114] Its amino acid sequence is known to be identical to a protein with erythroid-potentiating activity.[111,115] TIMP-1 produced by human bone marrow stromal cells has been shown in fact to stimulate colony formation by erythroid precursors.[116] The inhibitor has also been shown to have potent growth-promoting activity towards a range of normal and malignant cells.[114]

TIMP-2

Function

TIMP-2 binds preferentially to the latent and activated forms of MMP-2 but has also been shown to form a complex with activated MMP-1 and was reported to inhibit the activities of MMP-1, rat stromelysin and PUMP-1.[58,103,117-119] The inhibitor regulates not only the activity but also the processing of the proenzymes[118,119] by inhibiting the conversion of the intermediate forms to mature enzymes. The formation of a complex with interstitial procollagenase can only occur once a conformational change has been induced in the enzyme by organomercurials or plasmin. Complex formation in turn, inhibits the auto-proteolytic activation of the procollagenase.[119] Similarly, the interaction of TIMP-2 with the carboxyl-terminus of MMP-2 inhibits the plasma membrane-dependent activation of the enzyme and its conversion to a lower molecular weight active form.[89]

Expression

The gene coding for human TIMP-2 was localized to chromosome 17q25.[120] A full-length cDNA for TIMP-2 has been cloned and sequenced[110] and two distinct transcripts of 1 and 3.5 kb were detected in normal and malignant cells.[110] The 21 kDa protein shares 41% identity with human TIMP-1.[117] The positions of the 12 cysteine residues are conserved as are three out of four tryptophan residues[117] suggesting that these residues are critical for TIMP function. The secreted protein has 194 a.a. residues and unlike TIMP-1 is not glycosylated.[110] Although similar in composition and structure, the two proteins are immunogenically distinct as antibodies raised against TIMP-1 did not react with TIMP-2.[103,110]

Regulation

TIMP-1 and 2 are independently regulated in vitro and in vivo.[110] In colorectal carcinoma specimens for example, the levels of TIMP-1 but not TIMP-2 mRNA transcripts were found to be elevated relative to the adjacent normal mucosa.[110] In contrast to TIMP-1 which can be up-regulated by TGFβ in normal and malignant cells, TIMP-2 transcripts are down-regulated by this mediator in tumor cells but not in normal fibroblasts.[110] Phorbol esters which were consistently found to induce transcription of TIMP-1,[47,110] only slightly increased TIMP-2 transcript levels in a normal fibroblast cell line and either failed to modulate or reduced mRNA levels in melanoma cells and fibrosarcoma cells, respectively.[110]

TIMP-3

Expression

A full length cDNA for TIMP-3 has recently been cloned from a human breast carcinoma cDNA library.[121] The open reading frame spans 633bp and codes for a protein of 211 a.a. residues with a predicted molecular mass of 24,117 Da.[121] The gene was localized to the q12.1-q13.2 region of human chromosome 22.[122] TIMP-3 shares a 40.1% aa sequence identity with TIMP-1 and 44.9% with TIMP-2. The highest degree of aa sequence identity (82.4%) was found with ChIMP-3, a tissue metalloproteinase inhibitor produced by transformed chicken embryo fibroblasts.[121,123] The inhibitor is expressed in normal (heart, kidney, lung, liver, pancreas, uterus and skeletal muscle) and pathological (breast carcinomas) tissues with the highest expression found in placenta.[121,122] Several mRNA species are usually present in samples examined for TIMP-3.[121,122] The precise function and mechanism of action of this inhibitor have not yet been elucidated.

REGULATION OF METALLOPROTEINASES GENE EXPRESSION

Recent studies have begun to provide insight into the molecular mechanisms regulating expression of the metalloproteinase genes[5,6] particularly of MMP-1, MMP-3 and the gelatinases (which will be discussed here in greater detail).

In culture, connective tissue cells such as human skin fibroblasts[41,64] and synovial cells[65,124] produce undetectable or low constitutive levels of interstitial collagenase and stromelysin-1. These levels can be upregulated in response to various growth factors such as EGF and PDGF,[45,64,65,125] cytokines such as IL-1, and TNFα,[46,65,124,126] chemical agents such as phorbol ester[60,65,124] or by cell transfection with oncogenes such as *ras*.[125] On the other hand expression can be inhibited by glucocorticoids such as dexamethasone,[64,65] by retinoic acid,[124] or by inhibitory cytokines such as TGFβ.[127] The basal promoter functions are transmitted via a TATA box[128] which was identified in the 5' flanking regions of human MMP-1,[35,129] stromelysins 1 and 2[58] and MMP-9 genes[101] (see Fig. 9.3). In the 5' flanking region of the human 72 kDa type IV collagenase gene, on the other hand, a non conventional TATA box consisting of the sequence TACATCT provides the basal promoter activity.[82] This sequence is located 26 base pairs upstream of the major transcriptional start site and is surrounded by GC-rich sequences (GC boxes) which are involved in the binding of the transcriptional factor Sp1.[78,82] In addition to this promoter region, other short sequence elements that specifically bind transcription factors have been identified upstream from the TATA box and they also respond to the different regulatory signals for gene expression.[6] Putative enhancer and repressor regions have also been identified in the 5' region of the MMP-2 gene[82] and a potential consensus sequence for binding of the transcriptional factor Ap-2 was localized to exon 1.[78]

The sequence motif 5'-CATGAGTCAG-3' located in the 5' region of the human and rabbit MMP-1 gene, 70 bp upstream of the initiation site was found to be important for the regulation of gene expression by phorbol esters (PMA)[35,129,130] cytokines,[124,131] retinoic acid[124] and oncogenes.[132] This sequence appears to act as a transcriptional enhancer for MMP-1[124,129] as well as for MMP-9[101] and stromelysins 1 and 2.[58] It is known as the TPA-responsive

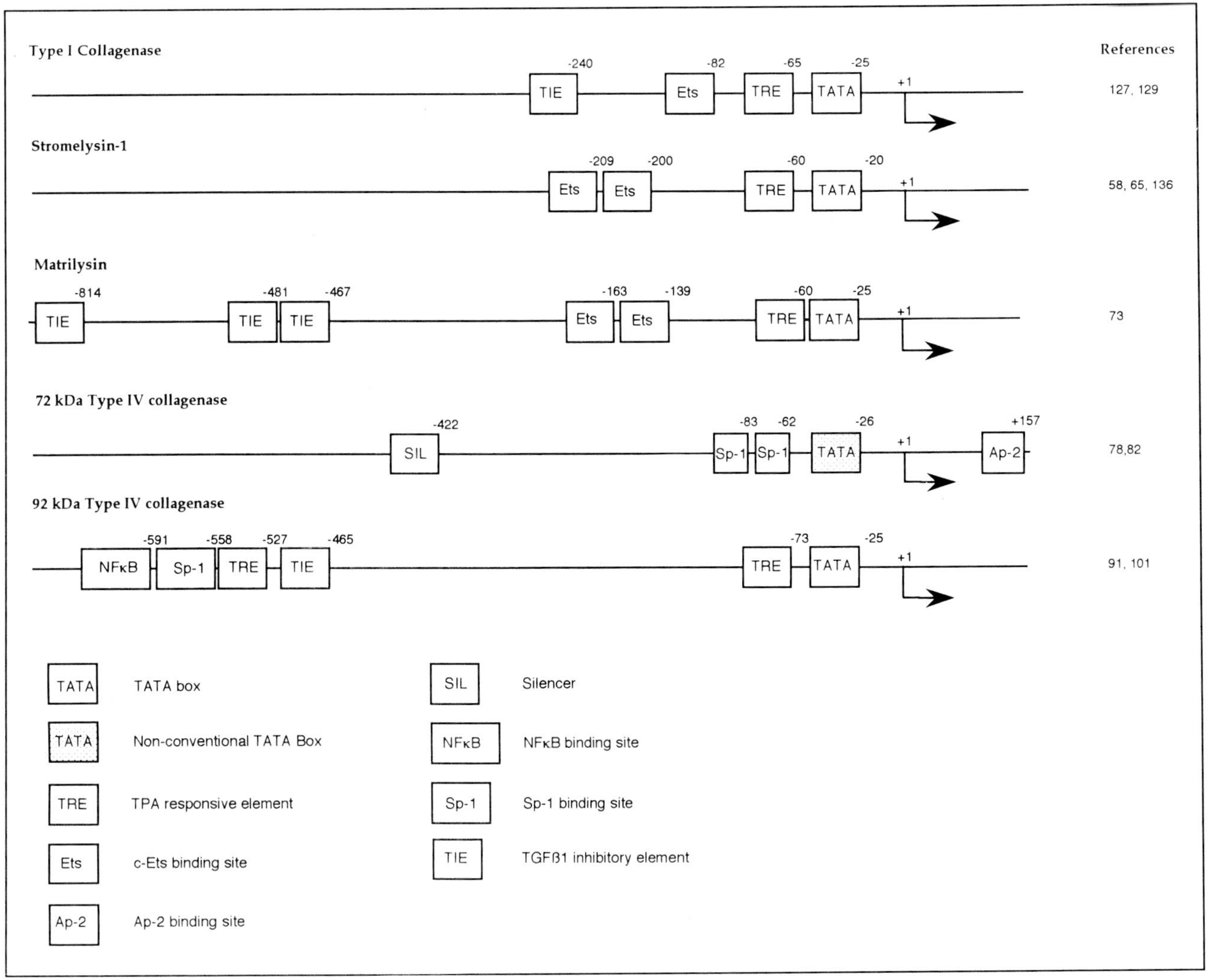

Fig. 9.3. Transcriptional elements involved in the regulation of metalloproteinases gene expression. The positions of the TATA box and of binding sites for transcriptional factors are indicated relative to the transcription initiation site (bent arrow).

element (TRE) or the Ap-1 sequence motif since it binds the Ap-1 (activator protein-1) nuclear transcription factor (see description below) in response to cell stimulation with 12-0-tetradecanoyl-phorbol-13-acetate (TPA).[129,130] This sequence is absent from the promoter region of MMP-2,[78] a factor which may be responsible for the differential regulation of this enzyme.

For maximal transcriptional activation of MMP-1 and MMP-7 in response to TPA, Ap-1 requires the cooperation of a PEA3-like element (which has sequence similarities to the polyoma virus enhancer and the ets motif).[73,133] It should be noted, however, that the requirements for a phorbol ester-induced response may differ depending on the cell system used. Thus for example, in normal rabbit fibroblasts, the upregulation of MMP-1 gene transcription in response to phorbol ester requires, in addition to the Ap-1 sequence[129,134] the cooperation of several transcriptional elements such as the PEA3-like sequence, a 5'-TTCA-3' motif and other elements present upstream of TRE.[129,134,135] Similarly, in normal human fibroblasts, maximal induction of the stromelysin promoter by IL-1 or phorbol ester requires cooperation between the Ap-1 sequence and several upstream regulatory sequences.[131] On the other hand, in tumor cells, the transfection of a recombinant construct expressing either the Ap-1 sequence alone or the Ap-1 and a PEA3-like sequence was sufficient to confer phorbol ester inducibility upon construct expression.[130]

Although the expression of MMP-9 can be induced by the same stimulants required for other MMPs, it appears to be regulated differently. Thus, the Ap-1 site although essential, is not sufficient for MMP-9 induction and the regulation of promoter activity with TPA requires the cooperative action of other upstream elements identified as NFκB- or Sp-1-like factors.[101] This was demonstrated by transfecting fibrosarcoma and hepatoma cells with a CAT (chloramphenicol acetyltransferase) reporter plasmid expressing variable size fragments of the MMP-9 5'-flanking region.

The mechanisms regulating TPA-induced transcription of stromelysins 1 and 2 are distinct. Expression of human stromelysin-1 can be significantly augmented by treatment with phorbol ester or growth factors such as EGF and cytokines such as IL-1 while the expression of stromelysin-2 is only minimally affected.[58,65] This difference in inducibility may be due to a base pair substitution in the fifth position of the AP-1 element in the human stromelysin-2 gene which results in a sequence distinct not only from stromelysin-1 but also from the consensus sequence expressed in all other metalloproteinase genes.[58] Studies on the rat stromelysin promoter have shown that it can be efficiently activated by the oncoproteins c-Ets-1 and -2.[6,136] This activation involves two DNA elements—a distal one which is a highly conserved palindrome composed of two binding sites for c-Ets (similar to the polyoma virus enhancer motif and also present in the human stromelysin promoter) and a proximal element which binds the AP-1 factor and is stimulated indirectly by c-Ets.[136] The c-Ets responsive element also mediates activation by the oncoproteins H-ras, v-src and v-mos. This may be the mechanism underlying the deregulated expression of stromelysin-1 in transformed and malignant cells.[136]

Protein kinase C (PKC) has been identified as a signaling intermediate for the activation pathway induced by the tumor promoter TPA as well as by growth factors and cytokines.[137-141] Stimulation of PKC triggers the interaction between the Ap-1 transcription factor and its AP-1 responsive element.[58,100,130,142] The Ap-1 transcription factor is a heterodimeric complex formed between the phosphoproteins encoded by the proto-oncogenes c-Jun and c-Fos.[6,143-145] This complex which contains multiple phosphorylation sites[137,145,146] binds to the Ap-1 recognition site regulating gene expression.[147] Protein dimerization during complex formation occurs through a coiled-coil interaction involving leucine zipper domains in each of the oncoproteins.[143,147,148] A DNA-binding domain is formed when regions rich in basic amino acids which are adjacent to the leucine zipper regions of each protein are brought into close proximity to each other. Genomic footprinting of the MMP-1 promoter has shown that TPA treatment results in the binding of Ap-1 to its DNA binding site in vivo.[149] The oncoprotein c-Jun, but not c-Fos, can form homodimers which can also bind the Ap-1 site but with lower affinity[147,148,150] suggesting that a Fos-independent mechanism for stimulation of the TRE enhancer activity may exist.[151] However, c-Fos expression was shown to be required for transcriptional activation of collagenase by TPA, IL-1 and transforming oncogene products[124,132] as transfection of synoviocytes with a construct expressing Fos DNA in the antisense orientation resulted in inhibition of MMP-1 stimulation by these agents[124] and an antisense Fos sequence could abolish transcription of constructs carrying the TRE sequence.[132] Moreover, in dermal fibroblasts, human recombinant IL-1β was shown to induce rapid increases in the expression of both c-Jun and c-Fos mRNA transcripts and a subsequent increase in procollagenase mRNA.[46] The requirement for c-Fos expression is thought to be related to the fact that the c-Jun-c-Fos heterodimer is significantly more stable than the c-Jun homodimer.[137,147]

The mechanism of Ap-1 activation by TPA is not completely understood. In some instances, stimulation of transcription through the TRE requires new protein synthesis as this indirect signal transduction pathway could be blocked by cycloheximide (an inhibitor of protein synthesis).[60] However, the level of Ap-1 binding and the transcriptional responses via the Ap-1 element can also be modulated in the absence of de novo protein synthesis suggesting that modification of pre-existing Ap-1 oncoproteins takes place probably through post-translational mechanisms such as phosphorylation and dephosphorylation of the Ap-1 proteins.[137] While in resting cells (human epithelial and fibroblastic cells), c-Jun is phosphorylated and incapable of efficient binding to the TRE, activation of PKC causes rapid dephosphorylation of c-Jun increasing Ap-1 binding activity.[137]

Also of importance in the stimulation of Ap-1 binding to its DNA sequence is Ref-1.[152] This 37 kDa nuclear protein is a redox component of the signal transduction pathway which acts through the conserved Cys residues located in the DNA-binding domains of Fos and Jun.[152] Oxidation of the Cys residue on Fos and Jun prevents DNA binding and Ref-1 can restore this activity probably by participating together with other factors in a redox cycle and serving as an electron donor for the Ap-1 proteins.

Ap-1 activity is also modulated by an inhibitory protein (IP-1). This protein was shown to specifically block DNA binding of Ap-1 derived from nuclear extracts or Fos/Jun proteins synthesized in vitro.[153] IP-1 activity is regulated by phosphorylation[153] as dephosphorylation restores the inhibitory activity whereas cAMP-dependent protein kinase A (PKA)-mediated phosphorylation abrogates function. The nature of the interaction between IP-1 and the oncoproteins is still a matter of speculation. The present evidence suggests that IP-1 does not compete with the Fos/Jun complex for the same DNA binding site but rather interacts loosely with the leucine zippers decreasing the dimerization potential of the Ap-1 proteins and thereby interfering with DNA binding.[153]

In addition to PKC-mediated signaling, other mechanisms may also be involved. Thus the stimulation of stromelysin-1 expression with PDGF, phosphatidyl-choline-hydrolyzing phospholipase C (PC-PLC) or the *ras* oncogene was shown to be independent of the PKC/TRE pathway[125] and an element located between bp-1303 and -754 of the stromelysin promoter region was found to be indispensable for activation via these mediators.[125]

TGFβ has been shown to inhibit the induction of interstitial collagenase and stromelysin-1 by growth factors.[48] The effect of TGFβ on expression of MMP-9 is, however, still unclear.[91,101] A TGFβ inhibitory element (TIE), a 10 bp motif required for the inhibitory effect,[127] has been identified in the promoter regions of several metalloproteinases such as MMP-1,[127] rat stromelysin,[127] MMP-9[101] and matrilysin.[73] The inhibition of gene expression appears to be mediated by the binding of a nuclear protein complex containing c-Fos to the TIE promoter region.[127]

Glucocorticoids are negative regulators of collagenase and stromelysin expression.[49,65,149,154] Dexamethasone for example, diminishes stromelysin induction by TPA, EGF and IL-1.[65] It was suggested that this may be mediated by direct binding of the hormone-receptor complex to its consensus element as three such sequences are located in the 5'-flanking region of the stromelysin gene.[65,155] Dexamethasone can also act at the post-transcriptional level and was shown to decrease the half-life of EGF-induced collagenase and stromelysin mRNA[64] without affecting EGF-induced transcription. Other indirect mechanisms of downregulation may also play a role. The glucocorticoid receptor-dexamethasone complex appears to interfere with the transcriptional activity of Ap-1.[49] The interaction of the steroid hormone-receptor complex with Ap-1 does not, however, abolish the ability of the latter to bind DNA.[149,156] There appears to be a mutual inhibitory interaction between the induced steroid hormone receptor and the Ap-1 complex[49,156,157] as the steroid receptor inhibits the Jun/Fos complex activity while the latter represses receptor activity and expression of hormone-dependent genes. The mutual inhibitory action is mediated by a protein-protein interaction[149,156] involving ligand binding and DNA-binding domains of the steroid hormone receptor and the leucine zipper of c-Jun[156] but does not involve their DNA response elements.

Tumor invasion has been shown to correlate with elevated extracellular levels of metalloproteinases. This may be linked to changes in regulatory mechanisms involving different intermediate molecules along the signal transduction pathway including upregulated expression of oncogenes, modulated expression of PKC or the increased autocrine expression of growth factors and cytokines which are positive regulators of metalloproteinase expression.[6,100,132,137] A better understanding of the signal transduction pathways regulating expression of this family of proteinases should have a direct impact on the design of new strategies for inhibition of invasion and metastasis.

REFERENCES

1. Yurchenco PD, Schittny JC. Molecular architecture of basement membranes. FASEB J 1990; 4:1577-90.
2. Tryggvason K, Hoyhtya M, Salo T. Proteolytic degradation of extracellular matrix in tumor invasion. Biochim Biophys Acta 1987; 907:191-217.
3. Mullins DE, Rohrlich ST. The role of proteinases in cellular invasiveness. Biochim Biophys Acta 1983; 695:177-214.
4. Mignatti P, Rifkin DB. Biology and biochemistry of proteinases in tumor invasion. Physiol Rev 1993; 73:161-95.
5. Matrisian LM. Metalloproteinases and their inhibitors in matrix remodeling. TIG 1990; 6:121-5.
6. Gutman A, Wasylyk B. Nuclear targets for transcription regulation by oncogenes. TIG 1991; 7:49-54.
7. Stetler-Stevenson WG, Liotta LA, Kleiner Jr DE. Extracellular matrix 6: role of matrix metalloproteinases in tumor invasion and metastasis. FASEB J 1993; 7:1434-41.
8. Stetler-Stevenson WG. Type IV collagenases in tumor invasion and metastasis. Canc Met Rev 1990; 9:289-303.
9. Woessner Jr JF. Matrix metalloproteinases and their inhibitors in connective tissue remodeling. FASEB J 1991; 5:2145-54.
10. Kleiner Jr DE , Stetler-Stevenson WG. Structural biochemistry and activation of matrix metalloproteases. Curr Opin Cell Biol 1993; 5:891-7.
11. Shapiro SD, Kobayashi DK, Ley TJ. Cloning and characterization of a unique elastolytic metalloproteinase produced by human alveolar macrophages. J Biol Chem 1993; 268:23824-9.
12. Sato H, Takino T, Okada Y et al. A matrix metalloproteinase expressed on the surface of invasive tumour cells. Nature 1994; 370:61-5.

13. Freije JMP, Diez-Itza I, Balbin M et al. Molecular cloning and expression of collagenase-3, a novel human matrix metalloprotinase produced by breast carcinoma. J Biol Chem 1994; 269:16766-73.
14. Sanchez-Lopez R, Alexander CM, Behrendtsen O et al. Role of zinc-binding- and hemopexin domain-encoded sequences in the substrate specificity of collagenase and stromelysin-2 as revealed by chimeric proteins. J Biol Chem 1993; 268:7238-47.
15. Docherty AJP, Murphy G. The tissue metalloproteinase family and the inhibitor TIMP: a study using cDNAs and recombinant proteins. Ann Rheum Dis 1990; 49:469-79.
16. Van Wart HE, Birkedal-Hansen H. The cysteine switch: a principle of regulation of metalloproteinase activity with potential applicability to the entire matrix metalloproteinase gene family. Proc Natl Acad Sci USA 1990; 87:5578-82.
17. Montgomery AMP, De Clerck YA, Langley KE et al. Melanoma-mediated dissolution of extracellular matrix: contribution of urokinase-dependent and metalloproteinase-dependent proteolytic pathways. Canc Res 1993; 53:693-700.
18. Mackay AR, Hartzler JL, Pelina MD et al. Studies on the ability of 65-kDa and 92-kDa tumor cell gelatinases to degrade type IV collagen. J Biol Chem 1990; 265:21929-34.
19. Lokeshwar BL, Selzer MG, Block NL et al. Secretion of matrix metalloproteinases and their inhibitors (tissue inhibitor of metalloproteinases) by human prostate in explant cultures: reduced tissue inhibitor of metalloproteinase secretion by malignant tissues. Canc Res 1993; 53:4493-8.
20. Urbanski SJ, Edwards DR, Maitland A et al. Expression of metalloproteinases and their inhibitors in primary pulmonary carcinomas. Br J Cancer 1992; 66:1188-94.
21. Azuma M, Tamatani T, Fukui K et al. Role of plasminogen activators, metalloproteinases and the tissue inhibitor of metalloproteinase-1 in the metastatic process of human salivary-gland adenocarcinoma cells. Int J Cancer 1993; 54:669-76.
22. Vassalli J-D, Pepper MS. Membrane proteases in focus. Nature 1994; 370:14-5.
23. Crabbe T, Smith B, O'Connell J et al. Human progelatinase A can be activated by matrilysin. FEBS 1994; 345:14-6.
24. Stetler-Stevenson WG, Krutzsch HC, Wacher MP et al. The activation of human type IV collagenase proenzyme. J Biol Chem 1989; 264:1353-6.
25. Springman EB, Angleton EL, Birkedal-Hansen H et al. Multiple modes of activation of latent human fibroblast collagenase: evidence for the role of a Cys^{73} active-site zinc complex in latency and a "cysteine switch" mechanism for activation. Proc Natl Acad Sci USA 1990; 87:364-8.
26. Woolley DE. Collagenolytic mechanisms in tumor cell invasion. Canc Met Rev 1984; 3:361-72.

27. Fields GB, Van Wart HE, Birkedal-Hansen H. Sequence specificity of human skin fibroblast collagenase. Evidence for the role of collagen structure in determining the collagenase cleavage site. J Biol Chem 1987; 262:6221-6.
28. Hughes C, Murphy G, Hardingham TE. Metalloproteinase digestion of cartilage proteoglycan. Biochem J 1991; 279:733-9.
29. Collier IE, Wilhelm SM, Eisen AZ et al. H-ras oncogene-transformed human bronchial epithelial cells (TBE-1) secrete a single metalloprotease capable of degrading basement membrane collagen. J Biol Chem 1988; 263:6579-87.
30. Crabbe T, O'Connell JP, Smith BJ et al. Reciprocated matrix metalloproteinase activation: a process performed by interstitial collagenase and progelatinase A. Biochem 1994; 33:14419-25.
31. Brinckerhoff CE, Ruby PL, Austin SD et al. Molecular cloning of human synovial cell collagenase and selection of a single gene from genomic DNA. J Clin Invest 1987; 79:542-6.
32. Goldberg GI, Wilhelm SM, Kronberger A et al. Human fibroblast collagenase. J Biol Chem 1986; 261:6600-5.
33. Templeton NS, Brown PD, Levy AT et al. Cloning and characterization of human tumor cell interstitial collagenase. Canc Res 1990; 50:5431-7.
34. Richards CD, Rafferty JA, Reynolds JJ et al. Procine collagenase from synovial fibroblasts: cDNA sequence and modulation of expression of RNA in vitro by various cytokines. Matrix 1991; 11:161-7.
35. Fini ME, Plucinska IM, Mayer AS et al. A gene for rabbit synovial cell collagenase: member of a family of metalloproteinases that degrade the connective tissue matrix. Biochem 1987; 26:6156-65.
36. Collier IE, Smith J, Kronberger A et al. The structure of the human skin fibroblast collagenase gene. J Biol Chem 1988; 263:10711-3.
37. Spurr NK, Gough AC, Gosden J et al. Restriction fragment length polymorphism analysis and assignment of the metalloproteinases stromelysin and collagenase to the long arm of chromosome 11. Genomics 1988; 2:119-27.
38. Moll UM, Lane B, Zucker S et al. Localization of collagenase at the basal plasma membrane of a human pancreatic carcinoma cell line. Canc Res 1990; 50:6995-7002.
39. Whitelock JM, O'Grady RL, Gibbins JR. Interstitial collagenase (Matrix Metalloproteinase 1) associated with the plasma membrane of both neoplastic and nonneoplastic cells. Inv Met 1991; 11:139-48.
40. Wilhelm SM, Eisen AZ, Teter M et al. Human fibroblast collagenase: Glycosylation and tissue-specific levels of enzyme synthesis. Proc Natl Acad Sci USA 1986; 83:3756-60.
41. He C, Wilhelm SM, Pentland AP et al. Tissue cooperation in a proteolytic cascade activating human interstitial collagenase. Proc Natl Acad Sci USA 1989; 86:2632-6.

42. Ishibashi M, Ito A, Sakyo K et al. Procollagenase activator produced by rabbit uterine cervical fibroblasts. Biochem J 1987; 241: 527-34.
43. Grant GA, Eisen AZ, Marmer BL et al. The activation of human skin fibroblast procollagenase. Sequence identification of the major conversion products. J Biol Chem 1987; 262:5886-9.
44. Quantin B, Murphy G, Breathnach R. Pump-1 cDNA codes for a protein with characteristics similar to those of classical collagenase family members. Biochem 1989; 28:5327-34.
45. McCarthy GM, Mitchell PG. Cheung HS. The mitogenic response to stimulation with basic calcium phosphate crystals is accompanied by induction and secretion of collagenase in human fibroblasts. Arth Rheu 1991; 34:1021-30.
46. Conca W, Kaplan PB, Krane SM. Increases in levels of procollagenase messenger RNA in cultured fibroblasts induced by human recombinant interleukin 1 or serum follow *c-jun* expression and are dependent on new protein synthesis. J Clin Invest 1989; 83:1753-7.
47. Murphy G, Reynolds JJ, Werb Z. Biosynthesis of tissue inhibitor of metalloproteinases by human fibroblasts in culture. Stimulation by 12-O-tetradecanoylphorbol-13-acetate and interleukin 1β in parallel with collagenase. J Biol Chem 1985; 260: 3079-83.
48. Edwards DR, Murphy G, Reynolds JJ et al. Transforming growth factor beta modulates the expression of collagenase and metalloproteinase inhibitor. EMBO J 1987; 6:1899-1904.
49. Jonat C, Rahmsdorf HJ, Park K-K et al. Antitumor promotion and antiinflammation: down-modulation of AP-1 (Fos/Jun) activity by glucocorticoid hormone. Cell 1990; 62:1189-1204.
50. Hasty KA, Jeffrey JJ, Hibbs MS et al. The collagen substrate specificity of human neutrophil collagenase. J Biol Chem 1987; 262: 10048-52.
51. Schnierer S, Kleine T, Gote T et al. The recombinant catalytic domain of human neutrophil collagenase lacks type I collagen substrate specificity. Biochem Biophys Res Commun 1993; 191:319-26.
52. Hasty KA, Pourmotabbed TF, Goldberg GI et al. Human neutrophil collagenase. A distinct gene product with homology to other matrix metalloproteinases. J Biol Chem 1990; 265:11421-4.
53. Busiek DF, Ross FP, McDonnell S et al. The matrix metalloprotease matrilysin (PUMP) is expressed in developing human mononuclear phagocytes. J Biol Chem 1992; 267:9087-92.
54. Wilhelm SM, Collier IE, Kronberger A et al. Human skin fibroblast stromelysin: structure, glycosylation, substrate specificity, and differential expression in normal and tumorigenic cells. Proc Natl Acad Sci USA 1987; 84:6725-9.
55. Koklitis PA, Murphy G, Sutton C et al. Purification of recombinant human prostromelysin. Biochem J 1991; 276:217-21.

56. Murphy G, Cockett MI, Stephens PE et al. Stromelysin is an activator of procollagenase. Biochem J 1987; 248:265-8.
57. Unemori EN, Bair MJ, Bauer EA et al. Stromelysin expression regulates collagenase activation in human fibroblasts. J Biol Chem 1991; 266:23477-82.
58. Sirum KL, Brinckerhoff CE. Cloning of the genes for human stromelysin and stromelysin 2: differential expression in rheumatoid synovial fibroblasts. Biochem 1989; 28:8691-8.
59. Saus J, Quinones S, Otani Y et al. The complete primary structure of human matrix metalloproteinase-3. J Biol Chem 1988; 263:6742-5.
60. Frisch SM, Clark EJ, Werb Z. Coordinate regulation of stromelysin and collagenase genes determined with cDNA probes. Proc Natl Acad Sci USA 1987; 84:2600-4.
61. Whitham SE, Murphy G, Angel P et al. Comparison of human stromelysin and collagenase by cloning and sequence analysis. Biochem J 1986; 240:913-6.
62. Sanchez-Lopez R, Nicholson R, Gesnel M-C et al. Structure-function relationships in the collagenase family member transin. J Biol Chem 1988; 263:11892-9.
63. Okada Y, Nagase H, Harris ED. A metalloproteinase from human rheumatoid synovial fibroblasts that digests connective tissue matrix components. Purification and characterization. J Biol Chem 1986; 261:14245-55.
64. Delany AM, Brinckerhoff CE. Post-transcriptional regulation of collagenase and stromelysin gene expression by epidermal growth factor and dexamethasone in cultured human fibroblasts. J Cell Biochem 1992; 50:400-10.
65. Frisch SM, Ruley HE. Transcription from the stromelysin promoter is induced by interleukin-1 and repressed by dexamethasone. J Biol Chem 1987; 262:16300-4.
66. Nicholson R, Murphy G, Breathnach R. Human and rat malignant-tumor-associated mRNAs encode stromelysin-like metalloproteinases. Biochem 1989; 28:5195-5203.
67. Muller D, Quantin B, Gesnel M-C et al. The collagenase gene family in humans consists of at least four members. Biochem J 1988; 253:187-92.
68. Basset P, Bellocq JP, Wolf C et al. A novel metalloproteinase gene specifically expressed in stromal cells of breast carcinomas. Nature 1990; 348:699-704.
68a. Murphy G, Segain JP, O'Shea M et al. The 28-kDa N-terminal domain of mouse stromelysin-3 has the general properties of a weak metalloproteinase. J Biol Chem 1993; 268:15435-41.
68b. Pei D, Majmudar G, Weiss SJ. Hydrolytic inactivation of a breast carcinoma cell-derived serpin by human stromelysin-3. J Biol Chem 1994; 269:25849-55.

69. Muller D, Wolf C, Abecassis J et al. Increased stromelysin 3 gene expression is associated with increased local invasiveness in head and neck squamous cell carcinomas. Canc Res 1993; 53:165-9.
70. Aaronson SA. Growth factors and cancer. Science 1991; 254:1146-53.
71. Lippman ME. The development of biological therapies for breast cancer. Science 1993; 259:631-2.
72. Marcotte PA, Kozan IM, Dorwin SA et al. The matrix metalloproteinase pump-1 catalyzes formation of low molecular weight (pro)urokinase in cultures of normal human kidney cells. J Biol Chem 1992; 267:13803-6.
73. Gaire M, Magbanua Z, McDonnell S et al. Structure and expression of the human gene for the matrix metalloproteinase matrilysin. J Biol Chem 1994; 269:2032-40.
74. Powell WC, Knox JD, Navre M et al. Expression of the metalloproteinase matrilysin in DU-145 cells increases their invasive potential in severe combined immunodeficient mice. Canc Res 1993; 53:417-22.
75. McDonnell S, Navre M, Coffey RJ Jr et al. Expression and localization of the matrix metalloproteinase pump-1 (MMP-7) in human gastric and colon carcinomas. Molec Carcin 1991; 4:527-33.
76. Murphy G, Cockett MI, Ward RV et al. Matrix metalloproteinase degradation of elastin, type IV collagen and proteoglycan. Biochem J 1991; 277:277-9.
77. Senior RM, Griffin GL, Fliszar CJ et al. Human 92- and 72-kilodalton type IV collagenases are elastases. J Biol Chem 1991; 266:7870-5.
78. Huhtala P, Chow LT, Tryggvason K. Structure of the human type IV collagenase gene. J Biol Chem 1990; 265:11077-82.
79. Collier IE, Bruns GAP, Goldberg GI et al. On the structure and chromosome location of the 72- and 92-kDa human type IV collagenase genes. Genomics 1991; 9:429-34.
80. Seftor REB, Seftor EA, Stetler-Stevenson WG et al. The 72 kDa type IV collagenase is modulated via differential expression of $\alpha_v\beta_3$ and $\alpha_5\beta_1$ integrins during human melanoma cell invasion. Canc Res 1993; 53:3411-5.
81. Kataoka H, DeCastro R, Zucker S et al. Tumor cell-derived collagenase-stimulatory factor increases expression of interstitial collagenase, stromelysin, and 72-kDa gelatinase. Canc Res 1993; 53:3154-8.
82. Templeton NS, Stetler-Stevenson, WG. Identification of a basal promoter for the human M_r 72,000 type IV collagenase gene and enhanced expression in a highly metastatic cell line. Canc Res 1991; 51:6190-3.
83. Stearns ME, Wang M. Type IV collagenase (M_r 72,000) expression in human prostate: benign and malignant tissue. Canc Res 1993; 53:878-83.

84. Levy AT, Cioce V, Sobel ME et al. Increased expression of the M_r 72,000 type IV collagenase in human colonic adenocarcinoma. Canc Res 1991; 51:439-44.
85. Chen J-M, Aimes RT, Ward GR et al. Isolation and characterization of a 70-kDa metalloprotease (gelatinase) that is elevated in Rous sarcoma virus-transformed chicken embryo fibroblasts. J Biol Chem 1991; 266:5113-21.
85a. Sorsa T, Salo T, Tyynela J et al. Tumor-associated trypsin (TAT)-2 converts 72kD progelatinase (MMP-2) to 62 kD gelatinase. Proc Am Assoc Canc Res 1995; 36:Abstr 614
86. Garbisa S, Pozzatti R, Muschel, RJ et al. Secretion of type IV collagenolytic protease and metastatic phenotype: induction by transfection with c-Ha-ras but not c-Ha-ras plus Ad2-E1a. Canc Res 1987; 47:1523-8.
87. Zucker S, Wieman JM, Lysik RM et al. Enrichment of collagen and gelatin degrading activities in the plasma membranes of human cancer cells. Canc Res 1987; 47:1608-14.
88. Emonard HP, Remacle AG, Noel AC et al. Tumor cell surface-associated binding site for the M_r 72,000 type IV collagenase. Canc Res 1992; 52:5845-8.
89. Strongin AY, Marmer BL, Grant GA et al. Plasma membrane-dependent activation of the 72-kDa type IV collagenase is prevented by complex formation with TIMP-2. J Biol Chem 1990; 268:14033-9.
90. Brown PD, Levy AT, Margulies IMK et al. Independent expression and cellular processing of M_r 72,000 type IV collagenase and interstitial collagenase in human tumorigenic cell lines. Canc Res 1990; 50:6184-91.
91. Huhtala P, Tuuttila A, Chow LT et al. Complete structure of the human gene for 92-kDa type IV collagenase. Divergent regulation of expression for the 92- and 72-kilodalton enzyme genes in HT-1080 cells. J Biol Chem 1991; 266:16485-90.
92. Seftor REB, Seftor EA, Gehlsen KR et al. Role of the $\alpha_v\beta_3$ integrin in human melanoma cell invasion. Proc Natl Acad Sci USA 1992; 89:1557-61.
93. Wilhelm SM, Collier IE, Marmer BL et al. SV40-transformed human lung fibroblasts secrete a 92-kDa type IV collagenase which is identical to that secreted by normal human macrophages. J Biol Chem 1989; 264:17213-21.
94. Mohtai M, Smith RL. Schurman DJ et al. Expression of 92-kD type IV collagenase/gelatinase (gelatinase B) in osteoarthritic cartilage and its induction in normal human articular cartilage by interleukin 1. J Clin Invest 1993; 92:179-85.
95. Librach CL, Werb Z, Fitzgerald ML et al. 92-kD type IV collagenase mediates invasion of human cytotrophoblasts. J Cell Biol 1991; 113:437-49.

96. Unemori EN, Hibbs MS, Amento EP. Constitutive expression of a 92-kD gelatinase (type V collagenase) by rheumatoid synovial fibroblasts and its induction in normal human fibroblasts by inflammatory cytokines. J Clin Invest 1991; 88:1656-62.
97. Bernhard EJ, Muschel RJ, Hughes EN. M_r 92,000 gelatinase release correlates with the metastatic phenotype in transformed rat embryo cells. Canc Res 1990; 50:3872-7.
98. Zucker S, Lysik RM, Zarrabi MH et al. M_r 92,000 type IV collagenase is increased in plasma of patients with colon cancer and breast cancer. Canc Res 1993; 53:140-6.
99. Goldberg GI, Strongin A, Collier IE et al. Interaction of 92-kDa type IV collagenase with the tissue inhibitor of metalloproteinases prevents dimerization, complex formation wtih interstitial collagenase, and activation of the proenzyme with stromelysin. J Biol Chem 1992; 267:4583-91.
100. Juarez J, Clayman G, Nakajima M et al. Role and regulation of expression of 92-kDa type-IV collagenase (MMP-9) in 2 invasive squamous-cell-carcinoma cell lines of the oral cavity. Int J Cancer 1993; 55:10-8.
101. Sato H, Seiki M. Regulatory mechanism of 92 kDa type IV collagenase gene expression which is associated with invasiveness of tumor cells. Oncogene 1993; 8:395-405.
102. Murphy G, Cawston TE, Reynolds JJ. An inhibitor of collagenase from human amniotic fluid. Purification, characterization and action on metalloproteinases. Biochem J 1981; 195:167-70.
103. Goldberg GI, Marmer BL, Grant GA et al. Human 72-kilodalton type IV collagenase forms a complex with a tissue inhibitor of metalloproteases designated TIMP-2. Proc Natl Acad Sci USA 1989; 86:8207-11.
104. Hahnel E, Harvey JM, Joyce R et al. Stromelysin-3 expression in breast cancer biopsies: clinico-pathological correlations. Int J Cancer 1993; 55:771-4.
105. Cooper TW, Eisen AZ, Stricklin GP et al. Platelet-derived collagenase inhibitor: characterization and subcellular location. Proc Natl Acad Sci USA 1985; 82:2779-83.
106. Welgus HG, Campbell EJ, Bar-Shavit Z et al. Human alveolar macrophages produce a fibroblast-like collagenase and collagenase inhibitor. J Clin Invest 1985; 76:219-24.
107. Carmichael DF, Sommer A, Thompson RC et al. Primary structure and cDNA cloning of human fibroblast collagenase inhibitor. Proc Natl Acad Sci USA 1986; 83:2407-11.
108. Herron GS, Banda MJ, Clark EJ et al. Secretion of metalloproteinases by stimulated capillary endothelial cells. J Biol Chem 1986; 261:2814-8.
109. Willard HF, Durfy SJ, Mahtani NN et al. Regional localization of the TIMP gene on the human X chromosome. Extension of a con-

served synteny and linkage group on proximal Xp. Hum Genet 1989; 81:234-8.
110. Stetler-Stevenson WG, Brown PD, Onisto M et al. Tissue inhibitor of metalloproteinases-2 (TIMP-2) mRNA expression in tumor cell lines and human tumor tissues. J Biol Chem 1990; 265:13933-8.
111. Docherty AJP, Lyons A, Smith BJ et al. Sequence of human tissue inhibitor of metalloproteinases and its identity to erythroid potentiating activity. Nature 1985; 318:66-9.
112. Schultz RM, Silberman S, Persky B et al. Inhibition by human recombinant tissue inhibitor of metalloproteinases of human amnion invasion and lung colonization by murine B16-F10 melanoma cells. Canc Res 1988; 48:5539-45.
113. Okada Y, Watanabe S, Nakanishi I et al. Inactivation of tissue inhibitor of metalloproteinases by neutrophil elastase and other serine proteinases. FEBS Lett 1988; 229:157-60.
114. Hayakawa T, Yamashita K, Tanzawa K et al. Growth-promoting activity of tissue inhibitor of metalloproteinases-1 (TIMP-1) for a wide range of cells. Fed Europ Biochem Soc 1992; 298:29-32.
115. Gasson JC, Golde DW, Kaufman SE et al. Molecular characterization and expression of the gene encoding human erythroid-potentiating activity. Nature 1985; 315:768-71.
116. Hayakawa T, Yamashita K, Kishi J-I et al. Tissue inhibitor of metalloproteinases from human bone marrow stromal cell line KM 102 has erythroid-potentiating activity, suggesting its possibly bifunctional role in the hematopoietic microenvironment. FEBS Lett 1990; 268:125-8.
117. Stetler-Stevenson WG, Krutzsch HC, Liotta LA. Tissue inhibitor of metalloproteinase (TIMP-2). J Biol Chem 1989; 264:17374-8.
118. Umenishi F, Umeda M, Miyazaki K. Efficient purification of TIMP-2 from culture medium conditioned by human hepatoma cell line, and its inhibitory effects on metalloproteinases and in vitro tumor invasion. J Biochem 1991; 110:189-95.
119. DeClerck YA, Yean TD, Lu HS et al. Inhibition of autoproteolytic activation of interstitial procollagenase by recombinant metalloproteinase inhibitor MI/TIMP-2. J Biol Chem 1991; 266:3893-9.
120. DeClerck Y, Szpirer C, Aly MS et al. The gene for tissue inhibitor of metalloproteinases-2 is localized on human chromosome arm 17q25. Genomics 1992; 14:782-4.
121. Uria JA, Ferrando AA, Velasco G et al. Structure and expression in breast tumors of human TIMP-3, a new member of the metalloproteinase inhibitor family. Canc Res 1994; 54:2091-4.
122. Apte SS, Mattei M-G, Olsen BR. Cloning of the cDNA encoding human tissue inhibitor of metalloproteinases-3 (TIMP-3) and mapping of the TIMP3 gene to chromosome 22. Genomics 1994; 19:86-90.

123. Pavloff N, Staskus PW, Kishnani NS et al. A new inhibitor of metalloproteinases from chicken: ChIMP-3. A third member of the TIMP family. J Biol Chem 1992; 267:17321-6.
124. Lafyatis R, Kim S-J, Angel P et al. Interleukin-1 stimulates and all-trans-retinoic acid inhibits collagenase gene expression through its 5' activator protein-1-binding site. Mol Endo 1990; 4:973-80.
125. Diaz-Meco MT, Quinones S, Municio MM et al. Protein kinase C-independent expression of stromelysin by platelet-derived growth factor, *ras* oncogene, and phosphatidylcholine-hydrolyzing phospholipase C. J Biol Chem 1991, 266:22597-602.
126. Brenner DA, O'Hara M, Angel P et al. Prolonged activation of jun and collagenase genes by tumor necrosis factor-α. Nature 1989; 337:661-3.
127. Kerr LD, Miller DB, Matrisian LM. TGF-β1 inhibition of transin/stromelysin gene expression is mediated through a Fos binding sequence. Cell 1990, 61:267-78.
128. Roeder RG. The complexities of eucaryotic transcription initiation: regulation of preinitiation complex assembley. TIBS 1991; 16:402-8.
129. Angel P, Baumann I, Stein B et al. 12-*O*-tetradecanoyl-phorbol-13-acetate induction of the human collagenase gene is mediated by an inducible enhancer element located in the 5'-flanking region. Molec Cell Biol 1987; 7:2256-66.
130. Angel P, Imagawa M, Chiu R et al. Phorbol ester-inducible genes contain a common *cis* element recognized by a TPA-modulated *trans*-acting factor. Cell 1987; 49:729-39.
131. Sirum-Connolly K, Brinckerhoff CE. Interleukin-1 or phorbol induction of the stromelysin promoter requires an element that cooperates with AP-1. Nuc Acids Res 1991; 19:335-41.
132. Schonthal A, Herrlich P, Rahmsdorf HJ et al. Requirement for *fos* gene expression in the transcriptional activation of collagenase by other oncogenes and phorbol esters. Cell 1988; 54:325-34.
133. Gutman A, Wasylyk B. The collagenase gene promoter contains a TPA and oncogene-responsive unit encompassing the PEA3 and AP-1 binding sites. EMBO J 1990; 9:2241-6.
134. Chamberlain SH, Hemmer RM, Brinckerhoff CE. Novel phorbol ester response region in the collagenase promoter binds Fos and Jun. J Cell Biochem 1993; 52:337-51.
135. Auble DT, Brinckerhoff CE. The AP-1 sequence is necessary but not sufficient for phorbol induction of collagenase in fibroblasts. Biochem 1991; 30:4629-35.
136. Wasylyk C, Gutman A, Nicholson R et al. The c-Ets oncoprotein activates the stromelysin promoter through the same elements as several non-nuclear oncoproteins. EMBO J 1991; 10:1127-34.
137. Boyle WJ, Smeal T, Defize LHK et al. Activation of protein kinase C decreases phosphorylation of c-Jun at sites that negatively regulate its DNA-binding activity. Cell 1991; 64:573-84.

138. Blumberg PM. Protein kinase C as the receptor for the phorbol ester tumor promoters: Sixth Rhoads Memorial Award Lecture. Canc Res 1988; 48:1-8.
139. Rosoff PM, Savage N, Dinarello CA. Interleukin-1 stimulates diacylglycerol production in T-lymphocytes by a novel mechanism. Cell 1988; 54:73-81.
140. Didier M, Aussel C, Pelassy C et al. IL-1 signaling for IL-2 production in T-cells involves a rise in phosphotidylserine synthesis. J Immunol 1988; 141:3078-80.
141. Nigg EA. Mechanisms of signal transduction to the cell nucleus. Adv Canc Res 1990; 55:271-310.
142. Lee W. Mitchell P, Tjian R. Purified transcription factor AP-1 interacts with TPA-inducible enhancer elements. Cell 1987; 49:741-52.
143. Curran T, Franza BR. Fos and Jun: the AP-1 connection. Cell 1988; 55:395-7.
144. Vogt PK, Bos TJ. jun: oncogene and transcription factor. Adv Canc Res 1990; 55:1-35.
145. Distel RJ, Spiegelman BM. Protooncogene c-fos as a transcription factor. Adv Canc Res 1990; 55:37-55.
146. Barber JR, Verma IM. Modification of *fos* proteins: phosphorylation of *c-fos*, but not *v-fos*, is stimulated by 12-tetradecanoyl-phorbol-13-acetate and serum. Mol Cell Biol 1987; 7:2201-11.
147. Turner R, Tjian R. Leucine repeats and an adjacent DNA binding domain mediate the formation of functional cFos-cJun heterodimers. Science 1989; 243:1689-94.
148. Gentz R, Rauscher III FJ, Abate C et al. Parallel association of Fos and Jun leucine zippers juxtaposes DNA binding domains. Science 1989; 243:1695-9.
149. Konig H, Ponta H, Rahmsdorf HJ et al. Interference between pathway-specific transcription factors: glucocorticoids antagonize phorbol ester-induced AP-1 activity without altering AP-1 site occupation in vivo. EMBO J 1992; 11:2241-6.
150. Halazonetis TD, Georgopoulos K, Greenberg ME et al. *c-jun* dimerizes with itself and with *c-fos*, forming complexes of different DNA binding affinities. Cell 1988; 55:917-24.
151. Kerr LD, Holt JT, Matrisian LM. Growth factors regulate transin gene expression by *c-fos*-dependent and *c-fos*-independent pathways. Science 1988; 242:1424-7.
152. Xanthoudakis S, Curran T. Identification and characterization of Ref-1, a nuclear protein that facilitates AP-1 DNA-binding activity. EMBO J 1992; 11:653-65.
153. Auwerx J, Sassone-Corsi P. IP-1: a dominant inhibitor of Fos/Jun whose activity is modulated by phosphorylation. Cell 1991; 64:983-93.

154. Schule R, Evans RM. Cross-coupling of signal transduction pathways: zinc finger meets leucine zipper. TIG 1991; 7:377-81.
155. Quinones S, Saus J, Otani Y et al. Transcriptional regulation of human stromelysin. J Biol Chem 1989; 264: 8339-44.
156. Schule R, Rangarajan P, Kliewer S et al. Functional antagonism between oncoprotein c-Jun and the glucocorticoid receptor. Cell 1990; 62:1217-26.
157. Yang-Yen H-F, Chambard J-C, Sun Y-L et al. Transcriptional interference between c-Jun and the glucocorticoid receptor: mutual inhibition of DNA binding due to direct protein-protein interaction. Cell 1990; 62:1205-15.

CHAPTER 10

Lysosomal Endoproteases

John S. Mort and Anneliese D. Recklies

INTRODUCTION

Increasing evidence indicates that lysosomal proteinases play a role in tumor invasion and metastasis, as well as in other normal and pathological conditions involving connective tissue degradation. The primary role of these enzymes is generally believed to be intracellular protein turnover where the combined action of the lysosomal endopeptidases and exopeptidases, in the specialized low pH environment of the lysosome, reduces proteins to their constituent amino acids for reutilization. However, extracellular forms of the lysosomal proteinases have been described in many pathological situations. In this review we will describe the properties of these enzymes stressing aspects related to their extracellular role.

PROTEINASE CLASSES

Since the 1960s, proteinases have been grouped according to their catalytic mechanisms leading to the establishment of four proteinase classes.[1] Three of these are named based on the residues critical for, and characteristic of, the catalytic mechanism. These are the serine, cysteine and aspartic proteinases. In the case of the fourth class, the metalloproteinases, a metal ion, in most cases Zn^{2+}, is the crucial component. Class membership can usually be determined by using proteinase inhibitors specific for each catalytic mechanism. The lysosomal proteinases found in most cells

Cell Adhesion and Invasion in Cancer Metastasis, edited by Pnina Brodt.

are members of the cysteine (cathepsins B, H, L, S and others) and aspartic (cathepsin D) classes while specialized cells such as the polymorphonuclear leukocyte and the mast cell contain lysosome-like granules with unique serine proteinases (cathepsin G, elastase, chymases and tryptases). The main mediators of normal extracellular matrix turnover, the matrix metalloproteinases (for example collagenase, stromelysin and the gelatinases) are generally not stored intracellularly but are secreted directly following their synthesis. Lysosomal endopeptidases which belong to the cysteine and aspartic proteinase classes are the subject of this review.

SYNTHESIS AND PROCESSING OF LYSOSOMAL PROTEINASES

Due to their potentially hazardous nature, all proteinases are synthesized as inactive, higher molecular weight precursors. The N-terminal signal peptide is followed by a proregion which acts as a potent inhibitor.[2] This region must be removed to generate active enzyme. Lysosomal proteinase precursors are glycoproteins which are tagged for transport to the lysosome by substitution of their N-linked oligosaccharides with mannose-6-phosphate. Receptors in the endosomes of the *trans* Golgi network then bind this modified sugar and transport the precursor to the lysosome. As the vesicle contents acidify, the enzyme is released from the receptor which is then recycled. Two mannose-6-phosphate receptors have been described, one being cation dependent, the other cation independent. The former mediates intracellular transport to the lysosome, while the later is involved in re-uptake of secreted precursor forms. The cation-independent mannose-6-phosphate receptor is identical to the insulin-like growth factor II (IGF-II) receptor.[3] In addition this sorting process appears to be complemented by other receptors which bind to the lysosomal proteinase precursors in a mannose 6-phosphate independent manner.[4] Concomitant with transport, acidification also triggers activation of the enzyme by proteolytic cleavage of the propeptide, in most cases apparently by autoprocessing. Once in the lysosome, further processing can occur to generate multiple chain forms of the enzymes (Fig. 10.1).

Unscheduled proteolytic activity of lysosomal cysteine proteinases is restricted by two mechanisms. A series of inhibitors (the cystatins) are present both in the cytoplasm and in the extracellular

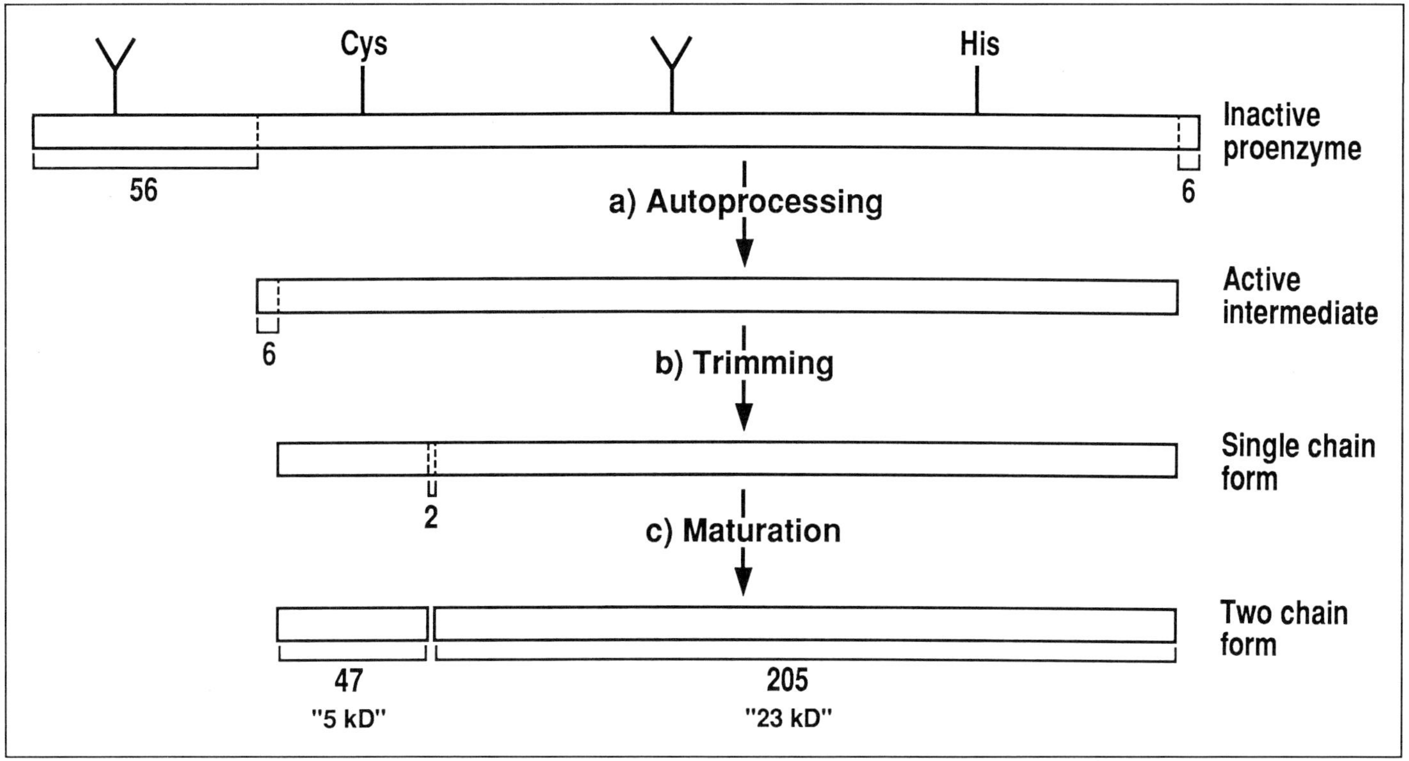

Fig. 10.1. Steps in the processing of cathepsin B. Schematic depiction of the procathepsin B polypeptide chain showing the position of N-linked oligosaccharides (Y), the catalytic cysteine and histidine residues and the various processing steps elucidated based on in vitro *and* in vivo *studies.*[38,49] *The number of residues removed at each step and the size of the final product are indicated. (a) On acidification, autoprocessing is initiated whereby the propeptide is cleaved and dissociates from the enzyme yielding the active proteinase. A six residue C-terminal extension is also removed by the peptidyl dipeptidase activity of cathepsin B. (b) The N-terminal extension is then removed by the successive actions of the lysosomal proteinase dipeptidyl peptidase IV to yield the single chain lysosomal form. (c) A later event (probably a prelude to enzyme turnover) is conversion to the two chain form, most likely due to the action of cathepsin L, with removal of a dipeptide.*

milieu. In addition, the mature forms of the cysteine proteinases, cathepsin B, cathepsin H and cathepsin L have a built-in "self-destruct" mechanism whereby they undergo spontaneous denaturation when they are exposed to neutral to alkaline pH on release from the acidic environment of the lysosome.[5]

THE CYSTEINE PROTEINASES

The cysteine proteinases present in the lysosome are all members of the papain family, named after papain—an exhaustively studied proteinase isolated from the papaya plant. Amino acid residues critical for peptide bond hydrolysis are absolutely conserved in all members of this family and there is a high level of homology in regions around these residues.[6] Other parts of the protein show more sequence diversity. Detailed sequence analysis of the proteinases of the papain family has lead to further division of its members into two sub-families.[7] Cathepsins L, H and S are most similar to the founder member, papain. They have long propeptides (~100 residues) and the length of the mature protein is approximately the same (~200 residues). Cathepsin B is distinct, as it contains a shorter propeptide (62 residues) but has a larger sized mature protein.

Following the initial proposal of Schechter and Berger in 1967,[8] based on studies with different synthetic substrates, cysteine proteinases appear to have binding sites for a maximum of seven amino acid residues—usually four N-terminal to, and three C-terminal to the cleavage position. These have been designated the S_4 to S_1 and S_1' to S_3' subsites on the enzyme which accept the P_4 to P_1 and P_1' to P_3' substrate residues, respectively. The substrate specificity of proteinases is of great interest since it provides an important guide for inhibitor design. The unprimed sites in the cysteine proteinases have been reasonably well characterized. In most cases the P_2 position is the most dominant with most cysteine proteinases preferring a hydrophobic residue such as Phe in this position. Cathepsin B is an interesting exception since it will also accept a basic residue such as Arg in its S_2 subsite as an alternative to a large hydrophobic residue. While different members of the cysteine proteinase family have slightly different preferences at their other subsites no clear rules for specificity have been established. This lack of absolute specificity makes the use of synthetic substrates for assay of a individual cysteine proteinase in the presence of other members of the family difficult unless care is taken to allow for the differential contribution of the other enzymes. Generally these proteinases can be assayed using dipeptide-based substrates having a blocked N-terminus and a chromogenic or fluorogenic leaving group. Good examples are carbobenzyloxy-phenylalanyl-arginine-*p*-nitroanilide (Z-Phe-Arg-pNA) and carbobenzyloxy-argininyl-arginine-4-methyl-7-coumarylamide (Z-Arg-Arg-MCA). In the former case the Phe and Arg residues occupy the P_2 and P_1 positions, respectively. Since most studies have been carried out using such substrates there has been much less characterization of prime site specificity although this is a fruitful area for exploration.

CATHEPSIN L

Cathepsin L is a potent, 25 kDa endopeptidase capable of degrading proteoglycan, collagen and elastin.[9-11] It can be assayed using the synthetic substrates described above and has a preference for Phe in the P_2 position. The mature form of the enzyme is extremely susceptible to denaturation at neutral to alkaline pH. Large amounts of procathepsin L, originally termed MEP (major excreted protein) are secreted by some tumor cells[12] and its synthesis is stimulated by phorbol esters. An active stable form of cathepsin L is found in organ cultures of rat mammary gland.[13] Re-uptake of extracellular procathepsin L is less efficient compared to other proforms of lysosomal enzymes due to a lower degree of N-linked oligosaccharide substitution.[14]

CATHEPSIN S

Cathepsin S is a recently characterized cysteine proteinase originally isolated from the spleen. It is a potent endopeptidase capable of degrading collagen and elastin.[15,16] Unique amongst the lysosomal cysteine proteinases characterized so far, cathepsin S is stable at alkaline pH conditions.[17] Recent studies using recombinant cathepsin S have demonstrated the value of the synthetic substrate Z-Val-Val-Arg-MCA for activity measurement of this enzyme.[17] A role for cathepsin S in tumor invasion has not yet been established.

CATHEPSIN H

While having some endopeptidase activity, cathepsin H appears to be primarily an aminopeptidase. It is often assayed with substrates such as Arg-MCA although is has the ability to hydrolyze the same endopeptidase substrates as are used for the other cysteine proteinases. Like cathepsins L and B, the mature form is unstable at neutral pH. Although cathepsin H has been detected extracellularly, its involvement in extracellular matrix degradation has not been established.

CATHEPSIN B

Of the many lysosomal cysteine proteinases so far characterized cathepsin B has received the most attention. It is unique in that it functions both as an endopeptidase and peptidyldipeptidase (removing dipeptides from the C-terminus of protein or peptide substrates). The three dimensional structure of the mature 30 kDa protein revealed the structural basis for this dual activity.[18] When compared to the stereotypical cysteine proteinases such as papain, cathepsin B was found to have an additional structural element containing a strategically positioned histidine residue that can act as an acceptor for the free negative charge on the carboxy terminus present on the exopeptidase substrate. Cathepsin B can be quantitated with relative accuracy using synthetic substrates due to its ability to accept an Arg residue in the P_2 position. The basis for this specificity has been elucidated at the molecular level.[19,20]

Using synthetic substrates, cathepsin B was found to be active over a wide pH range.[21] However, activity in the neutral to alkaline range is severely limited by the instability of the enzyme. The enzyme has been shown to degrade various proteins present in the extracellular matrix, such as proteoglycan, fibronectin and collagen.[9,10,22]

A role for extracellular cathepsin B in tumor invasion has been recognized for almost two decades.[23-25] Both procathepsin B and an active form stable at neutral pH have been found in tumor cell and organ cultures.[26] In addition, elevated levels of cathepsin B have been associated with the metastatic phenotype.[27] More recently a cell membrane-associated form of cathepsin B has been described[28] and an alternatively spliced cathepsin B mRNA lacking exon 2 has been reported which codes for a form of the enzyme lacking the signal peptide and seven residues of the propeptide.[29] Expression of this truncated mRNA would certainly modify the intracellular targeting of the proteinase.

OTHER CYSTEINE PROTEINASES

In addition to the well characterized enzymes outlined above, for which protein and/or cDNA sequences have been determined, several other activities have been described over the years which may be due to the action of other, as yet incompletely characterized cysteine proteinases. During the last year, two new members of the cysteine proteinase family were cloned and sequenced, one from osteoclasts termed cathepsin K,[30] and a second, designated cathepsin O, from a breast carcinoma.[31] Both of these are closely related to cathepsin L. It remains to be seen how many other members of this family are as yet only partially characterized or perhaps completely undescribed at present and to what extent other enzymes contribute in activity assays previously believed to be specifically measuring the well characterized members of the cysteine proteinase family.

CATHEPSIN D

The most abundant proteinase in the lysosome is usually the aspartic proteinase cathepsin D. The mature enzyme is a 34 kDa proteinase with broad specificity. Although stable at neutral pH, its extracellular role is however severely limited due to the very low pH range over which it is active. The extensive work of Rochefort's group, following up the observation of the release of procathepsin D from the MCF-7 human breast tumor cell line on stimulation with estrogen,[32] has demonstrated a positive correlation of total cathepsin D levels in primary breast tumor cytosols and the incidence of metastasis, leading to its use as a prognostic tool for disease-free intervals and overall survival.[33,34] Cathepsin D appears to be overexpressed in the tumor cells, but not in surrounding stromal cells,[35] and this might be due to the presence of a mixed promoter containing an estrogen-responsive element in the cathepsin D gene.[36] In experimental models it has been demonstrated that maturation of cathepsin D to its active form is necessary for increased incidence of metastasis,[37] implying a role for cathepsin D activity in disease progress.

EXTRACELLULAR FORMS

Both the proforms and active forms of lysosomal proteinases have been associated with tumor invasion and metastasis. It appears that under some circumstances the proenzymes are routed into the secretory pathway rather than to the lysosome. Heterologous expression of recombinant cysteine proteinase precursors has allowed detailed study of the activation mechanism of these enzymes and has provided an explanation for the existence of stable extracellular forms. In the cases of cathepsins B,[38,39] S[17] and L,[40] autoprocessing is initiated on acidification. This can occur by an intermolecular mechanism. There is also evidence for intramolecular processing. If cathepsin B is taken as an example the first step in autoprocessing is cleavage of the propeptide, leaving a six residue N-terminal extension relative to the ultimate lysosomal form. The region around this cleavage site appears to be exposed allowing cleavage by many other proteinases. Thus processing may not be restricted to the acidic conditions of the intracellular transport vesicles and lysosomes, but may also occur in the more neutral extracellular environment. In the lysosome the N-terminal extension is trimmed off, six residues are removed from the C-terminus and as a much later event a dipeptide is removed from the mature enzyme at a structurally exposed location[20] to yield a two chain form of the enzyme (Fig. 10.1).

The cleaved propeptide by itself is an extremely potent, specific inhibitor of cathepsin B[2]. However, binding is only effective at neutral pH while the complex dissociates on acidification, leading to degradation of the propeptide. Recent evidence indicates that complexes between the processed enzyme and the propeptide occur in organ culture.[41] Unlike the processed enzyme, the enzyme/propeptide complex is stable at neutral pH. Under the acidic conditions used in the assay of cathepsin B, dissociation and degradation of the propeptide occur allowing substrate hydrolysis to progress. Preliminary evidence suggests that the processing of cathepsin L proceeds by a similar mechanism.

The stability of partially processed cysteine proteinase intermediates in the extracellular milieu provides an effective mechanism for maintaining the normally unstable mature enzymes in a stable form until changes in conditions allow dissociation of the enzyme/inhibitor complex. Recent evidence indicates that tumor cells can acidify the extracellular matrix in their immediate vicinity.[42] This would lead to dissociation of the complex and liberation of active enzyme. In vitro and in vivo studies using cysteine proteinase inhibitors also indicate that these enzymes are involved in the invasive process[43-45] and that this role is dependent on the activity of these enzymes.

INHIBITORS

In addition to the general plasma proteinase inhibitor, α_2-macroglobulin, a family of specific cysteine proteinase inhibitors, the cystatins[46] is widely distributed intracellularly and in the extracellular fluid. No specific inhibitors have been reported for cathepsin D. Presumably they are not essential since this enzyme is only proteolytically active at the low pH found in the lysosome.

Substantial progress has been made in the development of specific low molecular weight inhibitors that could be used therapeutically. Screening of microorganisms yielded pepstatin, a potent general inhibitor of the aspartic proteinases, and E64 a potent irreversible inhibitor of cysteine proteinases of the papain family (Fig. 10.2). Modification of the peptidyl portion of E64 has allowed the introduction of selectivity for cathepsin B. Also by esterifying the charged groups on the inhibitor, prodrugs have been produced which pass through cell membranes allowing entry into the intracellular compartment. De-esterification then leads to intracellular generation of active inhibitor that allows differentiation between intracellular and extracellular proteinase activity. Several chemical strategies have been employed for the design of cysteine proteinase inhibitors and these have been used with some success in vivo. These include the fluoromethylketones[47] and acyloxymethylketones.[48] Rational design of specific, nontoxic cysteine proteinase inhibitors continues to be a very active field of investigation.

CONCLUSIONS

Our understanding of the role of lysosomal proteinases in connective tissue degradation continues to develop. Clearly lysosomal proteinase precursors are secreted from tumor cells and can undergo processing either on local acidification of the extracellular environment or through proteolytic cleavage by other proteinases. The identification of complexes between the cleaved propeptide and the active proteinase suggests a mechanism for stabilizing the extracellular enzyme. New information on the structure of mammalian cysteine proteinases and their complexes with inhibitors should aid in the rational design of new specific, bioactive proteinase inhibitors with ultimate therapeutic benefit.

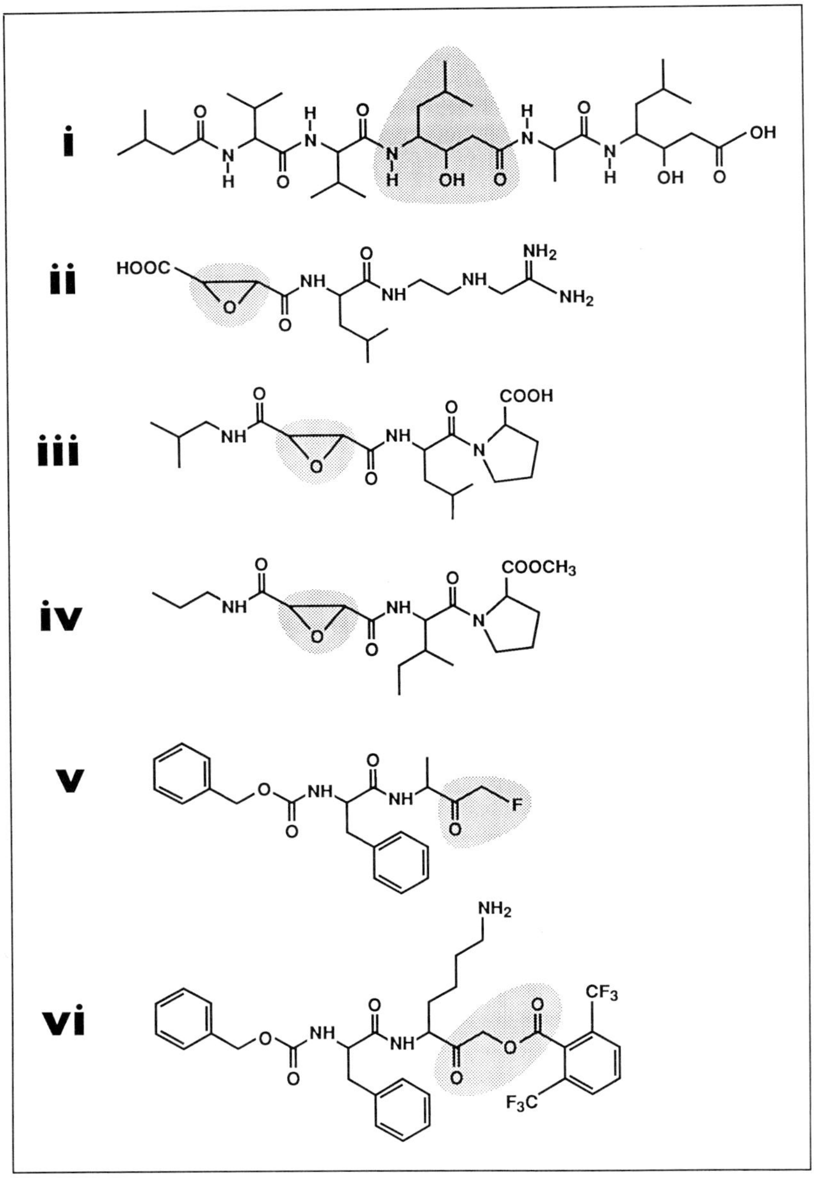

Fig. 10.2. Inhibitors of aspartic and cysteine proteinases. Structures of peptidyl inhibitors of aspartic (i) and cysteine (ii-vi) proteinases. In each case the shaded area indicates the moiety interacting with the active site of the enzyme. (i) Pepstatin, a potent inhibitor of cathepsin D showing the statine residue which acts as a transition-state analog. (ii) E64 [trans *epoxysuccinyl-leucylamido-(4-guanidino)butane] an irreversible inhibitor of cysteine proteinases of the papain family. The oxirane ring (shaded region) reacts with the active site cysteine residue. (iii) Isobutyl-*trans*-epoxysuccinyl-Leu-Pro, an E64 derivative which is highly selective for cathepsin B*[50] *(iv) CA076 methyl ester, a prodrug form of a variant of (iii) which has been shown to penetrate cells*[51] *(v) Carbobenzyloxy-Phe-Ala-fluoromethylketone which has been shown to inactivate cysteine proteinases* in vivo *in an arthritis model.*[52] *(vi) An example of the acyloxymethylketone family of cysteine proteinase inhibitors.*[53]

REFERENCES

1. Hartley BS. Proteolytic enzymes. An Rev Biochem 1960; 29:45-72.
2. Fox T, de Miguel E, Mort JS et al. Potent slow-binding inhibition of cathepsin B by its propeptide. Biochem 1992; 31:12571-6.
3. Kornfeld S, Mellman I. The biogenesis of lysosomes. Annu Rev Cell Biol 1989; 5:483-525.
4. McIntyre GF, Erickson AH. Procathepsins L and D are membrane-bound in acidic microsomal vesicles. J Biol Chem 1991; 266:15438-45.
5. Barrett AJ, Buttle DJ, Mason RW. Lysosomal cysteine proteinases. ISI Atlas Sci Biochem 1988:256-60.
6. Kamphuis IG, Drenth J, Baker EN. Thiol proteases. Comparative studies based on the high-resolution structures of papain and actinidin, and on amino acid sequence information for cathepsins B and H, and stem bromelain. J Mol Biol 1985; 182:317-29.
7. Karrer KM, Peiffer SL, Ditomas ME. Two distinct gene subfamilies within the family of cysteine protease genes. Proc Nat Acad Sci 1993; 90:3063-7.
8. Schechter I, Berger A. On the size of the active site in proteases. I. Papain. Biochem Biophys Res Commun 1967; 27:157-62.
9. Nguyen Q, Mort JS, Roughley PJ. Cartilage proteoglycan aggregate is degraded more extensively by cathepsin L than by cathepsin B. Biochem J 1990; 266:569-73.
10. Maciewicz RA, Wotton SF, Etherington DJ et al. Susceptibility of the cartilage collagens types II, IX and XI to degradation by the cysteine proteinases, cathepsins B and L. FEBS Lett 1990; 269:189-93.
11. Mason RW, Johnson DA, Barrett AJ et al. Elastinolytic activity of human cathepsin L. Biochem J 1986; 233:925-7.
12. Gal S, Gottesman MM. The major excreted protein of transformed fibroblasts is an activatable acid-protease. J Biol Chem 1986; 261:1760-5.
13. Recklies AD, Mort JS. Rat mammary gland in culture secretes a stable high molecular weight form of cathepsin L. Biochem Biophys Res Commun 1985; 131:402-7.
14. Dong J, Sahagian GG. Basis for low affinity binding of a lysosomal cysteine protease to the cation-independent mannose 6-phosphate receptor. J Biol Chem 1990; 265:4210-7.
15. Kirschke H, Wiederanders B, Brömme D et al. Cathepsin S from bovine spleen. Purification, distribution, intracellular localization and action on proteins. Biochem J 1989; 264:467-73.
16. Kirschke H, Schmidt I, Wiederanders B. Cathepsin S. The cysteine proteinase form bovine lymphoid tissue is distinct from cathepsin L (EC 3.4.22.15). Biochem J 1986; 240:455-9.
17. Brömme D, Bonneau PR, Lachance P et al. Functional expression of human cathepsin S in *Saccharomyces cerevisiae*. Purification and characterization of the recombinant enzyme. J Biol Chem 1993; 268:4832-8.

18. Musil D, Zucic D, Engh RA et al. The refined 2.15 Å x-ray crystal structure of human liver cathepsin B: the structural basis for its specificity. EMBO J 1991; 10:2321-30.
19. Hasnain S, Hirama T, Huber CP et al. Characterization of cathepsin B specificity by site-directed mutagenesis. The importance of Glu^{245} in the S_2-P_2 specificity for arginine and its role in transition state stabilization. J Biol Chem 1993; 268:235-40.
20. Jia Z, Hasnain S, Hirama T et al. Crystal structure of recombinant rat cathepsin B and a cathepsin B-inhibitor complex: implications for structure-based inhibitor design. J Biol Chem 1995; 270:5527-5533.
21. Hasnain S, Hirama T, Tam A et al. Characterization of recombinant rat cathepsin B and nonglycosylated mutants expressed in yeast. New insights into the pH-dependence of cathepsin B catalysed hydrolyses. J Biol Chem 1992; 267:4713-21.
22. Guinec N, Dalet-Fumeron V, Pagano M. "In vitro" study of basement membrane degradation by the cysteine proteinases, cathepsins B, B-like and L. Digestion of collagen IV, laminin, fibronectin, and relases of gelatinase activities from basement membrane fibronectin. Biol Chem Hoppe-Seyler 1993; 374:1135-46.
23. Poole AR, Tiltman KJ, Recklies AD et al. Differences in secretion of the proteinase cathepsin B at the edges of human breast carcinomas and fibroadenomas. Nature 1978; 273:545-7.
24. Recklies AD, Tiltman KJ, Stoker AM et al. Secretion of proteinases from malignant and nonmalignant human breast tissue. Cancer Res 1980; 40:550-6.
25. Recklies AD, Mort JS, Poole AR. Secretion of a thiol proteinase from mouse mammary carcinomas and its characterization. Cancer Res 1982; 42:1026-32.
26. Mort JS, Recklies AD. Interrelationship of active and latent secreted human cathepsin B precursors. Biochem J 1986; 233:57-63.
27. Sloane BF, Dunn JR, Honn KV. Lysosomal cathepsin B: correlation with metatstatic potential. Science 1981; 212:1151-3.
28. Sloane BF, Rozhin J, Johnson K et al. Cathepsin B: association with plasma membrane in metastatic tumors. Proc Nat Acad Sci 1986; 83:2483-7.
29. Gong Q, Chan SJ, Bajkowski AS et al. Characterization of the cathepsin B gene and multiple mRNAs in human tissues: evidence for alternative splicing of cathepsin B pre-mRNA. DNA Cell Biol 1993; 12:299-309.
30. Inaoka T, Bilbe G, Ishibashi O et al. Molecular cloning of human cDNA for cathepsin K: novel cysteine proteinase predominantly expressed in bone. Biochem Biophys Res Commun 1995; 206:89-96.
31. Velasco G, Ferrando AA, Puente XS et al. Human cathepsin O. Molecular cloning from a breast carcinoma, production of the ac-

tive enzyme in *Escherichia coli*, and expression analysis in human tissues. J Biol Chem 1994; 269:27136-42.
32. Capony F, Morisset M, Barrett AJ et al. Phosphorylation, glycosylation, and proteolytic activity of the 52-kD estrogen-induced protein secreted by MCF7 cells. J Cell Biol 1987; 104:253-62.
33. Spyratos F, Maudelonde T, Brouillet J-P et al. Cathepsin D: an independent prognostic factor for metastasis of breast cancer. Lancet 1989; 2:1115-8.
34. Duffy MJ, Reilly D, Brouillet JP et al. Cathepsin D concentration in breat cancer cytosols: correlation with disease-free interval and overall survival. Clin Chem 1992; 38:2114-6.
35. Zhao Y, Escot C, Maudelonde T et al. Correlation between mannose-6-phosphate/IGFII receptor and cathepsin D RNA levels by in situ hybridization in benign and malignant tumors. Cancer Res 1993; 53:2901-5.
36. Cavailles V, Augereau P, Rochefort H. Cathepsin D gene is controlled by a mixed promotor, and estrogens stimulate only TATA–dependent transcription in breast cancer cells. Proc Nat Acad Sci 1993; 90:203-7.
37. Liaudet E, Garcia M, Rochefort H. Cathepsin D maturation and its stimulatory effect on metastasis are prevented by addition of KDEL retention signal. Oncogene 1994; 9:1145-54.
38. Rowan AD, Mason P, Mach L et al. Rat procathepsin B. Proteolytic processing to the mature form *in vitro*. J Biol Chem 1992; 267:15993-9.
39. Mach L, Mort JS, Glössl J. Maturation of human cathepsin B. Proenzyme activation and proteolytic processing of the precursor to the mature proteinase, *in vitro*, are primarily unimolecular processes. J Biol Chem 1994; 269:13030-5.
40. Ishidoh K, Kominami E. Multi-step processing of procathepsin L in vitro. FEBS Letts 1994; 352:281-4.
41. Mach L, Mort JS, Glössl J. Non-covalent complexes between the lysosomal proteinase cathepsin B and its propeptide account for stable, extracellular, high molecular mass forms of the enzyme. J Biol Chem 1994; 269:13036-40.
42. Young PR, Spevacek SM. Substratum acidification and proteinase acctivation by murine B16F10 melanoma cultures. Biochem Biophys Acta 1993; 1182:69-74.
43. Yagel S, Warner AH, Nellans HN et al. Suppression by cathepsin L inhibitors of the invasion of amnion membranes by murine cancer cells. Cancer Res 1989; 49:3553-7.
44. Weiss RE, Liu B C-S, Ahlering T et al. Mechanisms of human bladder tumor invasion: role of protease cathepsin B. J Urol 1990; 144:798-04.
45. Redwood SM, Liu B C-S, Weiss RE et al. Abrogation of the invasion of human bladder tumor cells by using protease inhibitor(s). Cancer 1992; 69:1212-9.

46. Turk V, Bode W. The cystatins: protein inhibitors of cysteine proteinases. FEBS Lett 1991; 285:213-9.
47. Rasnick D. Synthesis of peptide fluoromethyl ketones and the inhibition of human cathepsin B. Anal Biochem 1985; 149:461-5.
48. Smith RA, Copp LJ, Coles PJ et al. New inhibitors of cysteine proteinases. Peptidyl acyloxymethyl ketones and the quiescent nucleofuge strategy. J Am Chem Soc 1988; 110:4429-31.
49. Rowan AD, Feng R, Konishi Y et al. Demonstration by electrospray mass spectrometry that the peptidyldipeptidase activity of cathepsin B is capable of rat cathepsin B *C*-terminal processing. Biochem J 1993; 294:923-7.
50. Gour-Salin BJ, Lachance P, Plouffe C et al. Epoxysuccinyl dipeptides as selective inhibitors of cathepsin B. J Med Chem 1993; 36:720-5.
51. Buttle DJ, Murata M, Knight CG et al. CA074 methyl ester: a proinhibitor for intracellular cathepsin B. Arch Biochem Biophys 1992; 299:377-80.
52. Ahmed NK, Martin LA, Mwatts L et al. Peptidyl fluoromethylketones as inhibitors of cathepsin B. Implication for treatment of rheumatoid arthritis. Biochem Pharmacol 1992; 44:1201-7.
53. Pliura DH, Bonaventura BJ, Smith RA et al. Comparative behaviour of calpain and cathepsin B toward peptidyl acyloxymethyl ketones, sulphonium methyl ketones and other potential inhibitors of cysteine proteinases. Biochem J 1992; 288:759-62.

PART II

ADHESION RECEPTORS AND PROTEOLYTIC MECHANISMS IN CANCER INVASION AND METASTASIS

CHAPTER 11

Adhesion Receptors and Proteolytic Mechanisms in Cancer Invasion and Metastasis

Pnina Brodt

GENERAL INTRODUCTION

The ability of malignant cells to detach from the primary site of growth and develop new foci in distant organs (i.e. metastasize) is still the main reason for the failure of current conventional therapy to cure cancer and the major cause for cancer-related mortality. In the past three decades major progress has been made in elucidating many of the cellular and molecular mechanisms which underlie this process. These studies revealed that the establishment of metastases is the final outcome of a series of complex and intricate interactions between the cancer cells and their rapidly changing microenvironment. This complexity is further compounded by the fact that the primary malignancies often consist of heterogeneous cell populations with diverse potentials to invade and metastasize.[1-3]

The metastatic process has traditionally been viewed as consisting of the following steps: (1) detachment of individual cells from the primary lesion (2) invasion of local stroma (3) entry of single cells or aggregates of tumor cells into blood vessels directly

Cell Adhesion and Invasion in Cancer Metastasis, edited by Pnina Brodt.

or via lymphatic channels (intravasation) (4) arrest in the vasculature followed by extravasation, and (5) invasion into the parenchyma of the target organ site with the resulting establishment of secondary lesions. The newly formed lesions can themselves become the source of disseminating cells which repeat this cycle giving rise to tertiary metastases (see excellent reviews in refs. 1-5 and Fig. 11.1).

When these steps in the progression of metastasis are examined closely, it becomes evident that they must entail changes in the adhesive properties of the tumor cells, as indeed numerous studies have confirmed.[6,7] While the disruption of cell-cell and cell-extracellular matrix (ECM) associations precedes and is essential for tumor cell detachment from the primary mass, (i.e. step 1),[8] increased tumor cell attachment to microvessel endothelial cells and the subendothelial matrix are known to be required for tumor cell intravasation and extravasation.[6,7,9] These adhesive interactions not only serve to anchor the cells to the target tissue but can also trigger or potentiate a host of cellular functions which are critical to metastasis including cell migration, the release of degradative enzymes leading to dissolution of the ECM[4,10,11] and the release of matrix-bound growth factors such as bFGF which can stimulate tumor cell growth as well as initiate the process of neovascularization.[4,6,12] Tumor cell adhesion to the ECM may also provide a growth signal for the tumor cells and regulate proteolysis.[4,11,13] In addition to the interactions with ECM proteins and the endothelium, tumor cell attachment to parenchymal cells of the target organ site has also been identified as an important element in the organ-colonizing potential of some metastatic cancer cells[6,14,15] and tumor cell interaction with circulating blood cells such as platelets and leukocytes had been shown to affect their metastatic potential.[16]

In this chapter, the contributions of different mediators of cell adhesion and proteolysis to tumor invasion and to the progression of metastasis are discussed and their linked and highly coordinated expression and functions outlined. While the molecular mechanisms which underlie various steps in this process may be similar and overlapping, the discussion for the sake of clarity, has been divided into two parts dealing with the molecular interactions which dominate at the primary site and those in the vasculature. The literature in this field is vast and no discussion of this nature can

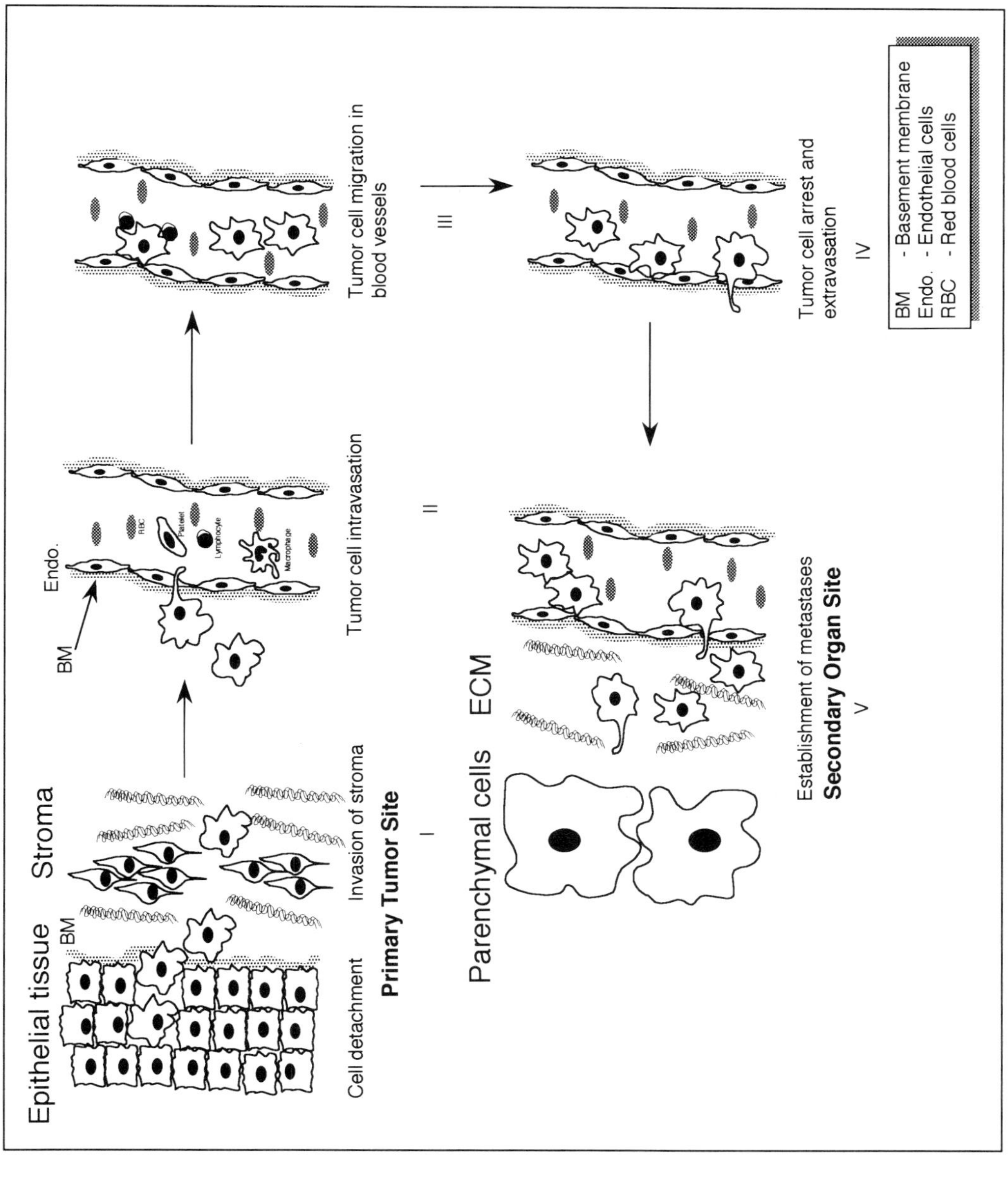

Fig. 11.1. The multi-step process of cancer metastasis.

be comprehensive. Omissions where they occurred were made necessary by space limitations and do not reflect a judgment. For additional information the reader is referred to the preceding chapters and to several recent excellent reviews.[1-3,12,17]

The major families of adhesion receptors and proteolytic enzymes which have been identified through the study of normal and pathological processes such as embryogenesis, wound healing, inflammation, immune responses and malignancy have been reviewed in the preceding chapters. Included among adhesion receptors are the integrins,[18,19] the Ig supergene family,[20] the Ca^{2+} dependent cadherins[21] and the selectins,[22] and among the proteinases are the metalloproteinases,[23] cysteine proteinases[24] and serine proteinases.[25] The role which members of each of these families play in the process of cancer dissemination will be discussed. Several other types of cell surface molecules such as CD44, N- and O-linked oligosaccharides and lectins have also been implicated in metastasis[7,12] and their functions will be briefly described.

DISRUPTION OF CELLULAR JUNCTIONS AND CELL MIGRATION AND INVASION AT THE PRIMARY SITE

THE ROLE OF ADHESION RECEPTORS

Cadherins

One of the earliest events signaling the initiation of an invasive-metastatic process is the loss of junctional contact between adjacent cells in the cancerous tissue. This is particularly true for cancers of epithelial origin where the normal tissue is characterized by a highly organized architecture maintained by tight intercellular junctions such as desmosomes or adherens junctions[26] (see chapter 3). Typically, as the epithelial cells acquire a malignant phenotype, the process is accompanied by a loss of this differentiated morphology, a reduction in the tight association with neighboring cells and the acquisition of a fibroblastic (mesenchymal) appearance.[27] In histological sections this is particularly evident in the invading front of the malignant tissue.[28] This loss of tissue cohesiveness which is essential for initiation of tumor cell migration into the adjacent stroma involves changes in the expression and function of several mediators of cellular communication. The best studied among these are E-cadherin and the integrins (see chapters 3 and 4).

E-cadherin is a 120 kDa transmembrane glycoprotein which is expressed on differentiated epithelium and mediates homotypic Ca^{2+} dependent cell-cell adhesion.[29] Numerous studies both in vitro and in vivo provide strong evidence that the alterations in E-cadherin expression and function constitute a major underlying mechanism in the acquisition of invasive properties by cells of epithelial origin. Thus, differentiated kidney epithelial cells (the MDCK line) were shown to acquire a fibroblast-like morphology and become invasive in collagen gels in response to treatment with a monoclonal antibody (MAb) to E-cadherin.[30,31] The same effect was seen when the cells were virally transformed and consequently lost E-cadherin expression.[31] Transfection of E-cadherin cDNA into transformed, invasive cells could reverse the invasive phenotype.[32,33] Moreover, in a series of carcinoma cell lines examined, an inverse correlation was noted between the degree of cellular differentiation and E-cadherin expression on one hand and cell invasion in collagen on the other.[21,27,32] In non-invasive cells, invasion could be enhanced by E-cadherin antisense RNA.[8] An inverse correlation between the levels of E-cadherin expressed on carcinoma cells and invasion has also been demonstrated in vivo.[31,34] Interestingly however, the metastatic potential did not consistently correlate with decreased E-cadherin levels but appeared to be associated with fluctuations in cadherin levels within a tumor cell population.[21,35,36]

Similar patterns were seen when biopsies derived from carcinomas of the breast, colon prostate and head and neck were analyzed. In these malignancies, an inverse correlation was generally observed between the degree of cellular differentiation and E-cadherin expression.[21,27,37] The incidence of metastasis correlated more often with reduced or heterogenous expression of E-cadherin in the primary lesion than with complete loss of expression, suggesting that some level of expression may have been necessary during the late stages of the metastatic process.[21,31]

In addition, recent studies suggest that E-cadherin function may be altered by mechanisms other than direct quantitative reduction in cell surface expression. Thus, they may be the outcome of structural modifications in the cadherin molecule or of changes in the expression and function of secondary molecules involved in cadherin-mediated signaling.[21,31] For example point mutations in the adhesion domain of the extracellular portion of the molecule have been demonstrated in invasive lobular carcinoma of the breast.[36]

Cadherin function is also dependent on the normal expression and function of α, β, and γ catenins and plakoglobin which bind to the intracellular domain of the cadherin molecule providing a link to the cytoskeleton[39] (see chapter 3). Recent studies have shown that in invasive tumor cell lines and tumor biopsies, which appeared to express normal levels of the relevant cadherin, catenin expression was reduced or lacking, resulting in a disfunctional adhesion system.[27,40,41] In addition, mutation or phosphorylation of catenin[42] have been associated with decreased cadherin function in human cancer cell lines and transformed cells, respectively.[42,43] Recently, an association between the cadherin-catenin complex and the phosphorylated 185 kDa gene product of the c-erb-2 oncogene has been reported and catenin was identified as the link molecule,[44] providing further support for a possible role of oncogene products in modulating the cadherin adhesion complex through phosphorylation of cadherin-associated proteins. This indirect mechanism for disruption of cadherin-mediated function may also be the basis for some of the known effects exerted by several growth factors on cell-cell adhesion and migration. Thus, several growth factors such as acidic fibroblast growth factor (aFGF) and hepatocyte growth factor (HGF scatter factor) have been shown to induce disruption of homotypic cell-cell adhesion.[45,46] HGF in particular has been implicated in the conversion of stationary well differentiated epithelial cells to migratory fibroblast-like cells—[27,46-48] a change reminiscent of the effects of downregulated expression of E-cadherin. The expression of these and other motility-inducing factors which bind to cell surface receptors with kinase activity such as PDGF and EGF is often upregulated in stromal cells associated with primary malignant lesions and the increased expression of their receptors has been noted in a variety of malignant cells.[27,49] This has given rise to speculation that the scattering effects of some of these factors may be mediated via phosphorylation of β catenin or other molecules acting downstream of the cadherin signaling pathway.[21,27,50] Recent studies have in fact shown that both HGF and EGF increase phosphorylation of β catenin and plakoglobin.[51]

Integrins

The disruption of cellular junctions at the primary site, while it is essential, cannot by itself account for the increased migration and invasive capacity of the cells detaching from the primary lesion.

In addition, the disruption of cell-extracellular matrix (ECM) associations, the acquisition of a motile phenotype and an increased proteolytic capability are necessary for the cells to degrade basement membrane and connective tissue barriers and migrate towards draining lymph or blood vessels. Cell-ECM adhesion is mediated mainly by members of the integrin family which consists of at least 21 distinct heterodimeric complexes. This family includes receptors for all major ECM proteins such as collagens, laminin, fibronectin and vitronectin as well as mediators of intercellular adhesion such as $\alpha_4\beta_1$ (VLA-4) and the integrins of the $\beta 2$ subfamily namely, LFA-1, MAC-1 and $\alpha_x\beta_2$ (see Table 4.1).

Studies on the role of integrins in the progression of malignant disease have relied on both descriptive and functional assays. They include immunohistochemical and biochemical analyses on cultured tumor cells with divergent invasive and metastatic potentials,[7,19,52] immunohistochemistry on tumor biopsies and the respective benign tissue[53] and functional adhesion and invasion assays using specific antibodies, inhibitory peptides and gene transfer technology to modify the cells (see also chapter 2).[7,19,54,55]

A review of the literature on integrin expression in transformed and malignant cells shows that the effect of transformation on integrin biosynthesis and expression defies generalization and is both cell system- and integrin-specific. For example, in chemically transformed osteosarcoma cells, increased expression of integrins $\alpha_1\beta_1$, $\alpha_2\beta_1$,and $\alpha_6\beta_1$ (the collagen and laminin binding integrins) which correlated with increased invasiveness in vitro was reported, while no change was seen in levels of $\alpha_5\beta_1$ or $\alpha_3\beta_1$.[56] On the other hand in rodent or human fibroblasts transformed by oncogenic viruses or transfected with oncogenes, the major difference observed was a reduction in $\alpha_5\beta_1$ expression.[57,58] Similarly, immunohistochemical analyses of tumor biopsies did not reveal a consistent pattern of change in integrin expression and often did not correspond to studies on cultured cells. Thus, in several types of carcinomas including carcinoma of the breast, lung and colon, downregulated expression of integrin subunits involved in collagen and laminin binding including α_2, α_3 and α_6 has been observed,[19] although increased α_3 expression in breast cancer has also been reported.[59] In melanoma lesions on the other hand, the expression of α_2 was found to be upregulated.[60,61] Interestingly however, increases in integrin-mediated adhesion to, and migration on basement membrane proteins have frequently been noted in highly metastatic tumor cells and integrin-specific antibodies or binding peptides have been shown to block tumor metastasis, suggesting that integrins can enhance the metastatic potential.[7,19,62-64]

The apparent inconsistencies in these findings may be related to variations in the assay systems used or in the cell types and tissues analyzed. They may also be reconciled in a dynamic model of tumor progression which takes into account the changing microenvironment of a cancer cell as it advances from a transformed to a highly metastatic cell. Thus, while reduced adhesion of malignantly transformed cells resulting from downregulation of normally expressed integrins (such as $\alpha_5\beta_1$ or $\alpha_2\beta_1$) and disrupted assembly of the corresponding matrix proteins (i.e. fibronectin or collagen) may be required initially for detachment and motility at the primary site, the process of migration through stroma, intravasation and colonization of new sites may require increased tumor attachment to matrix proteins or other cells.[12,65] These transient alterations in cell adhesion could conceivably be achieved through upregulation of adhesion receptors in response to signals such as cytokines, present in the microenvironment of the invaded site. Alternatively, they may result from alteration in processing and/or post-translational modification (i.e. glycosylation) of the normally expressed receptors[66] and/or the selection of minor subpopulations which pre-exist in the primary neoplasm and express the required levels of the relevant adhesion molecules. Indeed stromal cytokines such as TGFβ and IL-1 have been shown to regulate integrin expression (see chapter 4) and heterogeneity of integrin expression within neoplastic lesions has been observed.[7] As the models used to study integrin function in malignancy are diverse and include a range of cell types "frozen" in particular stages of this spectrum of malignant phenotypes, the diversity in integrin expression is not entirely surprising. In this section the discussion will focus on the evidence that modulated expression of specific integrins is an early event in transformation normally associated with progression towards an invasive phenotype.

The most consistent correlations reported between the acquisition of a tumorigenic/metastatic phenotype on the one hand and the modified expression of specific integrins on the other have been the following: (a) the reduced expression and function of $\alpha_5\beta_1$, the major fibronectin receptor, as a consequence of malignant transformation; (b) the upregulated expression of the vitronectin receptor $\alpha_v\beta_3$ as a function of malignant progression in melanoma; and (c) the altered expression and/or distribution of integrin $\alpha_6\beta_4$ in carcinomas.[19,53,65] A reduced expression of $\alpha_5\beta_1$ has been noted in transformed fibroblasts and Chinese hamster ovary (CHO) cells, in K562 leukemia cells and in several human malignancies such as ductal carcinoma of the breast and carcinomas of the colon and pancreas.[19] Conversely, overexpression of α_5 in CHO cells and in a human colon cancer cell line caused a loss of malignant properties such as anchorage independent growth and the ability to form tumors in nude mice.[66] It is of relevance to note that in some of these systems, the levels of $\alpha_5\beta_1$ were found to correlate with the expression of FN on the cell surface, suggesting that the negative regulatory effect of this integrin on tumorigenesis may be related to its role in matrix assembly. Increased production of fibronectin due to overexpression of its receptor may increase adhesion and limit motility, thereby reducing invasion. Moreover as the ECM is a source of signals regulating cell growth and differentiation, its defective assembly in transformed cells could deregulate these processes.[18,52,67] Indeed disrupted matrix assembly and reduced expression of ECM proteins such as FN is a characteristic finding in malignantly-transformed cells.[19,52]

In addition to its downregulated expression, other changes in cell surface $\alpha_5\beta_1$ have also been observed. These include changes in receptor distribution on the cell surface, a loss of ability to localize to focal contacts and changes in post-translational glycosylation.[68,69] Other alterations in the receptor such as oncogene-mediated phosphorylation of tyrosine on the cytoplasmic domain of the β subunit may also result in reduced ligand binding and in disrupted signal transduction by this integrin.[70] An alternative approach to interpreting these findings is that the disrupted expression and/or function of $\alpha_5\beta_1$ may, in fact, be a manifestation of the release of malignantly-transformed cells from cell adhesion-dependent growth signals. Thus whereas normal cells grow in an anchorage-dependent manner and depend on cell adhesion mechanisms for amplification of growth factor-derived mitogenic signals,[67,70] the acquisition by transformed cells of growth autonomy through constitutively activated or overexpressed oncogenes or through deregulated expression of suppressor genes, may decrease their requirement for matrix-derived survival and growth signals.[67]

In this respect, it should also be noted that the reduced function of $\alpha_5\beta_1$ is not a universal finding in all tumor models. In melanoma cells, ligation of the $\alpha_5\beta_1$ receptor by a fibronectin fragment was found to increase tumor cell proliferation.[71] While this may be related to differences in ligand availability on the different tumor cell types,[19] it is probably also an indication that the cellular context in which the integrins are expressed could influence their ultimate function and their contribution to the tumorigenic potential.[72]

The role of $\alpha_5\beta_1$ in melanoma cells may also be related to the increased expression in this malignancy of a second integrin namely, the vitronectin receptor $\alpha_v\beta_3$. This integrin has been identified as a marker of melanoma progression and its upregulated expression was shown to coincide with, and signal the conversion of "radial growth phase" (essentially benign) lesions to the more invasive and more aggressive "vertical growth phase" melanomas.[73] When invasive and metastatic melanoma biopsies were compared by immunohistochemistry to normal melanocytes and non-invasive melanomas, a consistent upregulation of its β_3 expression corresponding to increased malignancy was observed.[73] Several studies with established melanoma cell lines confirmed the histological findings and showed that increased expression of this receptor correlated with increased growth, tumorigenicity and metastasis of melanoma cells in vitro and in vivo.[74-76] Recently we have shown that human melanoma cells derived from lymph node metastases in nude mice had an increased adhesion to lymph node vitronectin which was mediated by $\alpha_v\beta_3$, implicating this receptor in metastasis to regional nodes.[77] Important work by Cheresh and associates revealed that this receptor plays a major role in regulating apoptosis (programmed cell death) and that its unhindered expression is required for melanoma cell survival and growth in 3-D interstitial collagen gels.[78] Furthermore, it was recently shown that upon proteolytic cleavage of native type I collagen, an RGD site is exposed, providing a ligand for receptor $\alpha_v\beta_3$.[79] As this collagen is the major ECM protein of the dermis, the increased cell surface expression of $\alpha_v\beta_3$ could provide the emerging melanoma cells with a survival advantage in their collagen-rich milieu. Moreover, integrin $\alpha_v\beta_3$ was implicated in the regulation of type IV collagenase expression in melanoma cells,[80] (a function which can apparently be mediated by $\alpha_5\beta_1$ when it predominates)[72] and linked to the expression, function and processing of the receptor for the urokinase plasminogen activator (uPA).[81] This vitronectin receptor can, therefore, regulate two major mechanisms of ECM proteolysis[82] (chapters 8 and 9) and thereby increase the invasive/metastatic potential of the melanoma cells.

Other mechanisms underlying malignant progression have also been linked, either directly or indirectly to the expression and function of $\alpha_v\beta_3$. Namely, malignant melanoma cells are known to acquire growth autonomy through upregulated synthesis and expression of growth factors and/or their receptors or through loss of susceptibility to growth arresting cytokines such as TGFβ and IL-6.[83] Among the factors which have been implicated in autocrine stimulation of melanoma cell growth is the type 1 insulin-like growth factor (IGF-1).[83,84] Recent studies have shown that insulin receptor substrate-1 (IRS-1), a signaling intermediate involved in binding Grb2 and PI-3 kinase, in response to insulin or IGF-1,[85] is an $\alpha_v\beta_3$-associated protein. This was also reported for the 190 kDa protein which is tyrosine-phosphorylated in response to PDGF.[86] This evidence points to an integration of the vitronectin receptor and growth factor-receptor induced signal transduction pathways with each amplifying the molecular responses activated by the other. In this respect it is noteworthy that in addition to malignant melanoma, receptor $\alpha_v\beta_3$ expression was also shown to be upregulated on other tumors of neural crest origin such as gliomas and neuroblastomas which are also growth regulated through autocrine production of IGF-1.[87,88] In addition the receptor was implicated in invasion and migration of epithelial malignancies such as pancreatic and colorectal carcinomas in which IGF-1 has been reported to play a role.[88-90]

When taken together the evidence on the role of $\alpha_v\beta_3$ in malignant melanoma may perhaps provide a paradigm for the critical role which integrins can play in the complex interactions between the emerging malignantly-transformed cells and their microenvironment. The following sequence of events may be postulated: During the pre-invasive, vertical growth phase, stromal cells or infiltrating leukocytes may respond to the localized melanocytic hyperplasia by releasing cytokines such as TGFβ, IL-6 or IL-1.[91] As the melanocytes progress towards a more malignant phenotype due to further carcinogenic insult (e.g. exposure to u.v. or chemicals),[92,93] and the deregulated expression of oncogenes and/or loss of suppressor genes these cytokines may contribute to the deregulation of ECM protein synthesis and assembly, and to upregulation of integrins such as $\alpha_v\beta_3$.[94] Increased attachment to vitronectin in the ECM through the upregulated receptor may in turn, tilt the balance in favor of cell migration because it can trigger both an increase in synthesis and clustering of the uPA receptors at focal adhesion contacts[81,95,96] and an increase in type IV collagenase production.[80] This net increase in proteolytic activity is likely to result in increased cell motility not only due to dissociation of cell-ECM contacts, but also because fragments of degraded ECM proteins can provide a chemotactic trigger.[6,97] In addition, ECM degradation may lead to the release and/or activation of ECM-bound growth factors such as bFGF, TGFβ and HGF/SF[98] and the ligation of $\alpha_v\beta_3$ may amplify signal transduction mechanisms induced through the binding of autocrine and/or paracrine IGF-1 to its receptor.[85] The combined effect of these molecular interactions is to further enhance melanoma cell proliferation leading to further amplification of this chain of events. Receptor $\alpha_v\beta_3$-mediated ligand binding may also result in a net increase in melanoma cell growth through its more direct role as a negative regulator of apoptosis.[78,99] Increased expression of $\alpha_v\beta_3$ on the melanoma cells my therefore contribute to increased cell survival, proliferation, migration and invasion, as well as to induction of angiogenesis through factors such as bFGF (see schematic representation in Fig. 11.2).

While the positive correlation between $\alpha_v\beta_3$ expression and melanoma malignancy is supported by most studies,[82,100] there are conflicting reports on the role of integrin VLA-4 ($\alpha_4\beta_1$) in the progression of this disease. In several recent analyses of human malignant melanoma specimens or cell lines, a positive correlation was found between expression of $\alpha_4\beta_1$ and the metastatic potential.[101-103] Similar observations were made with other malignancies such as renal cell carcinoma.[104] On the basis of in vitro, studies this increased metastatic potential was attributed to the increased adhesion of $\alpha_4\beta_1$ cells to cytokine-inducible VCAM-1 on the vascular endothelium—as this interaction could increase tumor cell arrest in the vasculature and induce the expression and secretion of the basement membrane-degrading proteinase MMP-2.[105,106] In an apparent contradiction to these observations, a recent report by Qian et al has shown that invasion and spontaneous metastasis of

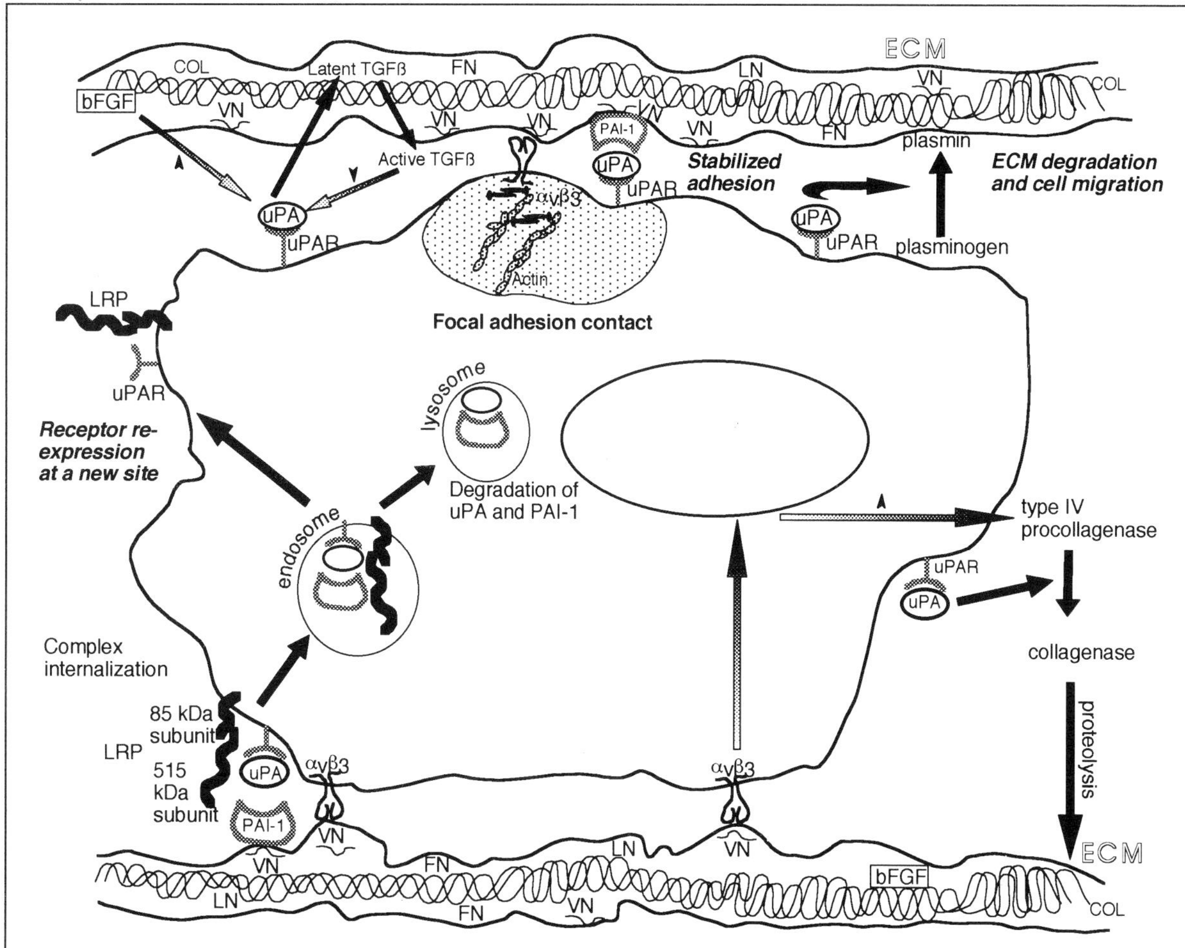

Fig. 11.2. The coordinated function and regulation of the vitronectin receptor $\alpha_v\beta_3$ and the urokinase receptor uPAR: revelance to metastasis. Shown is a schematic representation of the links between the vitronectic receptor and the uPAR at both the functional and transcriptional levels. Dotted arrows depict regulatory pathways, bold arrows depict protein processing and activation and arrow heads indicate up- or down- regulation. The abbreviations are: bFGF; basic fibroblast growth factor, COL: collagen, ECM; extracellular matrix, FN; fibronectic, LN; laminin, LRP; low density lipoprotein receptor related protein, PAI; plaminogen activator inhibitor, TGF; transforming growth factor, uPA; urokinase plasminogen activator, uPAR; urokinase plaminogen activator receptor, VN; vitronectin. Reprinted from: The role of integrin vitronectin receptor avb3 in melanoma metastasis. Cancer and Metastasis Reviews 1995; 14:341-352. Reproduced with permission from Kluwer Academic Publishers ©1995.

a murine B16 melanoma subline were suppressed by overexpression of $\alpha_4\beta_1$ through retroviral gene transfer.[107] This was attributed to increased $\alpha_4\beta_1$-mediated homotypic adhesion and a reduction in the cellular dissociation necessary for cell migration and invasion. Interestingly, no change was seen in the ability of the cells to colonize the lung in an experimental metastasis assay when the cells were inoculated iv. This suggests that the cellular interactions affected by overexpression of $\alpha_4\beta_1$ were playing a decisive role in the initial stages of dissemination, possibly by inhibiting cell detachment and invasion at the primary site, but were less critical during the latter stages when the tumor cells gained access to the circulation. The difference in these results may be related to interspecies differences, as a positive correlation of $\alpha_4\beta_1$ expression with metastasis was demonstrated mainly with human melanomas[101] while the inverse correlation was seen with an established murine melanoma line. Alternatively, they may suggest that while a moderate increase in the expression of this integrin can tilt the balance in favor of metastasis by providing a mechanism of tumor-endothelial recognition, overexpression as was achieved by gene transfer can result in immobilization of the cells through increased intercellular stickiness.

Another frequent observation has been a change in expression of laminin receptors, particularly the α_6-based integrins $\alpha_6\beta_1$ and $\alpha_6\beta_4$, in neoplasms of epithelial origin.[108] In the normal epithelium, receptor $\alpha_6\beta_4$ is generally localized in the basolateral aspects of the cells in areas of cell-ECM contact.[109,110] This orientation is apparently lost in malignant tissues and a more diffuse distribution of the receptor, which may be accompanied by either a reduced (e.g. breast or colon carcinomas) or increased (e.g. squamous carcinoma of the head and neck[111]) expression has been noted. In experimental skin carcinogenesis, a shift in the normal distribution of $\alpha_6\beta_4$ to suprabasilar expression was also found to be associated with increased risk for malignant progression.[112] The impact of such changes on the malignant phenotype[113] is not surprising in view of the important role played by α_6 integrins in adhesion and migration on laminin-containing matrices.[110,114] This is also demonstrated by results of a differential display analysis recently reported by Sager et al in which they identified the α_6 subunit as a potential tumor suppressor in breast cancer.[115] Since the heterodimers $\alpha_6\beta_1$ and $\alpha_6\beta_4$, although they utilize a common subunit, localize to different submembranous structures (i.e. focal contacts and hemidesmosomes, respectively)[109,110] and may mediate different post-adhesion events,[108,109,114,115] it is conceivable that cell behavior may ultimately be determined by the relative contribution of these integrins to laminin binding.

The patterns of expression observed for other integrins have been less consistent and appear to be tumor-type or model specific. For example, in one study the expression of $\alpha_3\beta_1$ in breast carcinoma tissue was reported to be increased and it was localized to areas of stromal-epithelial contact.[59] In contrast, decreased expression of α_3 and other VLA integrins was noted in a recent immunohistochemical study which compared primary breast carcinoma tissue to benign specimens[116] and reductions in α_3 expression were also observed in carcinomas of the colon and pancreas.[117] In a recent study, we found that overexpression of this integrin in a human rhabdomyosarcoma cell line suppressed both growth and tumorigenicity of these cells (Weitzman JB et al, manuscript submitted). Similarly, receptor $\alpha_2\beta_1$ was shown to be downregulated in colorectal carcinomas[118] but its overexpression in rhabdomyosarcoma cells resulted in increased metastatic potential.[114] These findings serve to reinforce the conclusion that while malignant transformation is invariably associated with changes in the expression patterns of different integrins, the contribution of particular integrins to the malignant phenotype may ultimately be dictated by the cellular context in which they are expressed.

More relevant perhaps to our understanding of the role of integrins in the initial stages of malignant transformation are questions pertaining not strictly to their levels but also to their state of activation, or to the expression and activation of downstream intermediates in their signal transduction pathways. Indeed, in recent studies, an upregulated expression of the focal adhesion kinase ($p125^{FAK}$) was noted in a large number of tumor specimens and this increase correlated with the invasiveness of the tumors.[119] As discussed above, this increased expression, if coupled with phosphorylation as a consequence of oncogenic transformation, may effectively release the cells from their dependence on integrin-mediated, post-adhesion signaling for essential cellular functions such as growth and migration.[67]

Taken together, the evidence described suggests that the molecular events underlying tumorigenesis are accompanied by profound changes in the expression and function of members of both the cadherin and integrin families and that these changes contribute to the disruption of normal communication between the emerging tumor cells and their immediate microenvironment. As recent evidence suggests that expression of integrins and cadherins may be coordinately regulated and their functions are linked,[120,121] it is possible that the changes observed in their expression in malignant cells are in fact related.

The Role of Nonintegrin Adhesion Molecules

In addition to receptors of the cadherin and integrin families, the modulated expression of other adhesion molecules may also contribute to tissue dissociation and cell motility at the primary site. These include the following: (1) the non-integrin 67 kDa-laminin receptor (or its 37 kDa precursor) whose expression on some malignant cells, including carcinoma cells of the colon, breast, ovaries, and liver was found to be upregulated and to correlate with the invasive potential of the cells,[108] (2) the galactoside-binding lectins (galectins) which interact with ECM laminin via its N-linked oligosaccharides and whose expression was found to be modulated on malignant cells,[108,122] and (3) the β-1,4 galactosyltransferase which has been implicated in the migration of a wide range of cells during physiological and pathological processes.[108,123] Collectively, the modified (up- or downregulated) expression of these molecules is thought to impact on the invasive and metastatic potentials of the tumor cells by modifying tumor cell interactions with the extracellular matrix. However, it is unclear at present whether they contribute to local invasiveness at the primary site, are involved in tumor cell adhesion to the subendothelial basement membrane during intravasation/extravasation or play a role in both stages of the metastatic cascade.[108]

Contributing to the adhesive/invasive phenotype of the malignant cells are also molecules which belong to the carcino-embryonic antigen (CEA) subgroup of the Ig supergene family[124,125] (see chapter 5). Members of this subfamily have been divided into two types, the glycophosphatidylinositol (GPI)-linked molecules which are normally upregulated on human malignancies and the transmembrane-linked molecules which appear to be downregulated in malignant cells.[126] While members of both of these groups mediate homotypic intercellular adhesion, their mode of adhesion as well as the molecular consequences of their expression and binding appear to be distinct.[126] Recent evidence derived from studies on the effect of ectopic expression of GPI-linked CEA in myoblasts suggest that these molecules can block cellular differentiation and thereby promote uninterrupted cell proliferation.[127] Furthermore, immunohistochemical analysis localized increased levels of this molecule to the invading front of colorectal carcinoma cells suggesting that it may also play a role in invasion at the primary site.[128] CEA molecules are also present in the serum and it has been suggested that in this soluble form they can block cell adhesion and therefore contribute to cell motility.[128,129] CEA molecules may also play an indirect role in cell motility by modulating the expression of other adhesion molecules such as integrins and cadherins.[130] In a recent study it was found that CEA can function as a ligand for galectin-3, a galactoside lectin widely expressed on colon carcinoma cells.[131]

Another adhesion molecule whose role in malignancy has received much attention in recent years is CD44. Initially identified as a homing receptor which mediates lymphocyte adhesion to high endothelial venules in the lymph node and regulates their maturation,[132] this molecule is now known to be expressed in multiple isoforms (splice variants) on a variety of cell types. In animal studies, one of these isoforms (CD44v) has been identified as a potential metastasis-inducing glycoprotein[133] and on various human malignancies such as carcinomas of the breast, colon and cervix, the upregulated expression of CD44 isoforms, particularly of splice variant CD44v6 has been associated with poor prognosis.[134-138] The role of these molecules in metastasis in vivo is not yet entirely clear. The most direct evidence for a link between this molecule and the invasive/metastatic phenotype was obtained in studies with a rat carcinoma model where over-expression of a variant form of CD44 was found to confer a metastatic potential on a non-metastatic tumor subline and an antibody to this variant blocked spontaneous lymph node metastasis (reviewed in refs. 140,141).[133,139]

CD44 isoforms are also known to be expressed in a highly restricted manner on other tissues. These include migrating epithelial cells at the basal layer of the epidermis and at the base of the intestinal crypts, cells in glandular tissue during prenatal development and lymphocytes during maturation induced by antigen-mediated activation.[142,143] This suggests that the expression of this molecule is associated with directed cell locomotion and transient states of proliferation.[142] CD44 is a receptor for hyaluronic acid, an ECM proteoglycan present on diverse cell types including endothelial and epithelial cells, leukocytes and fibroblasts.[144-146] It appears likely therefore that on tumor cells, it is involved in cell-ECM or cell-cell interactions which are inducive to tumor cell migration and/or growth.[141,145,147,148] Alternatively, it is also possible that variant forms of the molecule interfere with normal cell-ECM adhesion and thereby promote a more motile phenotype.[148]

When taken together these data suggest that the acquisition of a motile phenotype by malignantly-transformed cells—one of the first steps required in their progression towards a metastatic phenotype—is the cumulative result of the deregulated expression of multiple adhesion receptors which can occur as a consequence of internal (oncogenes) and or external (stromal factors) stimuli. As these changes are dynamic and none is likely to occur in isolation, a more complete understanding of the relative contribution which different adhesion mediators make to tumor invasiveness will be possible only when additional insight is gained into the reciprocal regulation among these receptors and the mechanisms of cross-talk between adhesion molecules and other growth modulators.

The Role of ECM-degrading Proteinases in Local Invasion

In addition to diminished cell-cell and cell-matrix contact through modulation of adhesion receptors, the detachment of malignant cells from the surrounding tissue and their migration through stroma also requires active proteolysis and degradation of otherwise impermeable extracellular matrix barriers.[13,149] Several families of proteinases have been implicated in this process including serine proteinases such as the urokinase plasminogen activator and elastase, metalloproteinases such as MMP-1, MMP-2, MMP-9 and others, and cysteine and aspartic lysosomal proteinases such as cathepsins L, B and D respectively[25] (Please see chapters 7-10 for comprehensive reviews on the structure and function regulation of members of each of family proteinase). These enzymes are also involved in proteolysis in a range of physiological processes such as wound healing, angiogenesis, trophoblast implantation, embryogenesis, morphogenesis and neurite outgrowth. In all of these processes, the proteolytic activity is tightly regulated at both the levels of synthesis and function. Enzyme activity is regulated at two levels a; the enzymes are produced and secreted in latent pro-enzyme forms and require activation through the coordinated action of other mediators. Second, natural inhibitors produced by the same or adjacent cells form a complex with the proteinases and block their activity.[150] The aggressive/invasive behavior of malignant cells is due to disruption of these regulatory mechanisms at either or both ends of the spectrum. i.e. enzyme expression may be constitutively upregulated and this may be accompanied by increased availability of activating enzymes and/or diminished production of endogenous inhibitors.[4,13]

There is ample evidence (direct and indirect) for the involvement of proteinases from all of these families in invasion at the primary site (as well as in the latter stages of metastasis as discussed below). Most of this evidence has been derived from three types of studies: (a) immunohistochemistry, immunochemistry and in situ hybridization which were performed on tumor biopsies provided evidence for the concentration of proteolytic enzymes at the invasive fronts of malignant lesions, as well as quantitative information linking the stage and aggressiveness of malignant disease to proteinase expression. (b) functional in vitro assays with different ECM models such as reconstituted or natural basement membranes and individual matrix proteins which were used to mimic tumor invasion, provided quantitative data on the invasive capabilities of tumor cells derived from dispersed specimens and cell lines and (c) animal models which provided evidence on the role of different proteinases in invasion and metastasis in vivo[151] (see chapter 7). A representative selection of these studies is described below. Together they demonstrate that members of each of the major families of proteinases can contribute to the invasive potential of the malignant cells through direct matrix degradation and/or through participation in proteolytic cascades.

The Role of the Metalloproteinases

The Gelatinases

Several metalloproteinases with the capacity to cleave collagen and other ECM proteins have been implicated in tumor invasion.[152] The evidence is particularly strong for the gelatinases MMP-2 and

MMP-9.[153] Immunoreactivity for MMP-2 has been demonstrated in a wide range of neoplasms including in gastric, breast, colon, ovarian and thyroid carcinomas and increased mRNA expression for both proteinases was found in several malignancies such as basal and squamous cell carcinomas of the skin. In these studies the normal adjacent tissues or respective benign hyperplasias showed little or no reactivity.[153-155] Upregulated expression of one or both of these enzymes which correlated with increased invasiveness in vitro and with metastatic potential in vivo was noted in various *ras*-transformed cells such as murine 3T3 fibroblasts, rat embryo fibroblasts or normal bronchial epithelial cells.[156-158] The invasiveness and metastatic potential of these and other highly metastatic cells could be blocked by overexpression, through gene transfer, of the tissue inhibitors of metalloproteinases TIMP-1 or TIMP-2, by exogenously added TIMPs or by antibodies specific to the metalloproteinases.[159-162]

It should be noted however that in cancerous lesions, expression of these enzymes may not be restricted to the malignant cells.[163] In a recent study, Gallegos et al have analyzed 40 colorectal carcinoma specimens and found that MMP-2 could be immunolocalized both to the ECM and to fibroblast-like cells surrounding the tumor cells, whereas MMP-9 was found mainly in infiltrating leukocytes.[164] These and other observations (see below) suggest that in vivo the host cell infiltrate can contribute significantly to ECM degradation and tumor cell invasion.

The expression of MMP-2 and MMP-9 can be regulated by different mediators. As discussed above, regulation of MMP-2 gene expression and tumor invasion by integrins $\alpha_v\beta_3$ and/or $\alpha_5\beta_1$ has been demonstrated in melanoma cells.[11,72] Inflammatory mediators such as TGFβ and IL-1,[149] which may be produced in vivo by reactive stroma and by infiltrating host leukocytes are known to regulate MMP-2 and MMP-9 expression respectively, and they provide therefore a mechanism for control of tumor cell-mediated proteolysis by the stroma. Furthermore, stromal cytokines such as TGFβ can also downregulate the expression of TIMP-2[165] thereby increasing the net level of proteolysis. As matrix degradation proceeds, the accumulation of laminin fragments and the changes in pH due to local tissue destruction[11,23,149,166] can also upregulate the expression of the gelatinases, particularly of MMP-2, resulting in further enhancement of invasion.

The proteolytic activity of MMP-2 and MMP-9 can also be regulated at the level of enzyme activation. While little is known about the natural activators of the latent forms of these metalloproteinases, in vitro studies suggest that other metalloproteinases (such as stromelysins or interstitial collagenase), as well as cysteine proteinases (i.e. cathepsins B and L) and uPA may mediate cleavage of the pro-gelatinases and trigger their activation.[149,150,152,167,168] These activating enzymes may be produced by the tumor cells or they may be contributed by reactive stromal cells or infiltrating leukocytes such as macrophages.[169,170] Recently, a membrane-bound metalloproteinase (MT-MMP) with an apparent progelatinase activation function has been identified.[171] This enzyme is expressed in malignant tissues and was also found in stromal cells of human colon, breast and head and neck carcinomas,[172] providing further evidence for the critical role that the stroma can play in regulating gelatinase-mediated proteolysis and invasion. Yet another factor in the complex stromal-epithelial interaction at the site of the emerging malignancy is the release, through matrix degradation, of stromal ECM-bound growth factors such as TGFβ and bFGF.[98] As some of these factors can modulate gelatinase expression, this may provide another invasion-amplifying mechanism in the reciprocal cascade of host-tumor interactions. This regulatory effect which the stroma can exert on the proteolytic capability of the tumor cells is thought to be the basis for the observation that the site of tumor implantation can influence the metastatic potential of tumor cells implanted into animals. In several studies it was found that tumors which were non-invasive and non-metastatic when implanted ectotopically (e.g. colorectal carcinoma implanted subcutaneously) acquired a metastatic potential when implanted in the orthotopic microenvironment.[173-175]

Interstitial Collagenase and the Stromelysins

Although less is known about the role of interstitial collagenase (MMP-1) in invasion at the primary site, its ability to degrade native types I II and III collagen which are the major constituents of connective tissues,[176] make it a likely participant in the process of local invasion. This is supported by experimental and clinical data. Thus, elevated levels of interstitial collagenase have been reported in the conditioned media or plasma membrane fractions of various tumors including naturally occurring murine mammary tumors,[177] invasive and metastatic rat carcinomas[178] and highly metastatic murine melanoma cells.[179] In several human malignancies such as colon carcinoma, small cell lung carcinoma and squamous cell carcinomas, increased enzyme production was seen in a large proportion of the malignant specimens but in some of these studies, enzyme production was localized to stromal fibroblastic cells in the malignant tissue.[180-182] Immunohistochemical analysis of a human pancreatic carcinoma line localized this enzyme to basal type microvilli of the plasma membrane where it would be optimally located for degradation of the stroma.[183] Functional studies using in vitro invasion models such as the chick embryo chorioallantoic membrane[184] and matrigel (Durko M et al, manuscript submitted) assays, confirmed that MMP-1 is involved in the degradation of these basement membrane-like matrices by human carcinoma and or melanoma cells, respectively. The evidence suggests that in these systems MMP-1 is engaged in a proteolytic cascade with other proteinases such as uPA, cathepsins and MMP-2.[13,184] This cooperative activity may involve processing of the latent pro-collagenase to its active form by another protease, or the activity of more than one enzyme may be required for complete degradation of the substrate. Indeed it was recently shown that MMP-1 and 2 are involved in a process of reciprocal activation.[185] Plasmin, the product of plasminogen activator proteolysis, was implicated in activation of collagenases[186] and the combined activity of MMP-1 and cathepsin B was shown to be required for complete proteolytic degradation of type X collagen.[187]

The synthesis of MMP-1 can be induced by various cytokines and growth factors such as IL-1, TNFα, PDGF, bFGF and EGF. These modulators can be released as part of the inflammatory exudate surrounding the malignant tissue or by activated stromal cells, providing another mechanism for enhancement of the invasive process by the stroma (see chapter 9).

Stromelysins-1 and -2 which degrade a broad spectrum of matrix proteins including fibronectin, laminin, types IV V and IX collagens, elastin and the protein core of proteoglycans[176,188,189] are also modulated by the same cytokines but less is known about their involvement in tumor invasion. While widely expressed in different types of connective tissue cells, there are conflicting reports regarding the expression of these metalloproteinases in tumor cells. For example, increased expression of stromelysin-2 mRNA was reported in epidermoid carcinomas, particularly in tumor cells located in proximity to disrupted basement membrane.[190] On the other hand, analysis of gastric and colonic carcinoma failed to show prominent expression of either stromelysin-1 or -2.[191] In another study, when expression of different MMPs in transformed rat embryo cells with divergent metastatic potentials was analyzed, the metastatic potential correlated with the expression of stromelysins-1 and -2 but not with the levels of the gelatinases MMP-2 and -9.[192] Furthermore, in a mouse skin carcinogenesis model, expression of stromelysin-1 was shown to correlate with tumor progression. In this model, expression was restricted to the stroma in the early stages of chemical carcinogenesis while the lesions were still essentially benign but an increased enzyme expression was seen in squamous carcinomas and highly metastatic spindle cell tumors.[193] This suggests that elevated expression of stromelysins in the stromal cells is an early event in tumor progression which precedes and may promote the acquisition of metastatic potential by the transformed epithelial cells. This may be due to the release of growth factors such as PDGF by the activated stroma,[98,194] or it may be related to the role that stromelysins can play in the activation of other metalloproteinases such as MMP-1 thereby contributing indirectly to the acquisition of an invasive capacity by the carcinoma cells.[195-197]

The evidence regarding the role stromelysin-3 in invasion is somewhat more conclusive. This enzyme was initially identified as a metalloproteinase specifically expressed by stromal cells in breast carcinoma tissue.[198] This has since been extended to include other types of cancer such as squamous cell carcinoma of the head and neck, basal cell carcinoma and carcinomas of the colon, lung and kidney,[197-202] but no expression has been observed in normal adult tissue.[202] In primary and metastatic lesions of colon carcinoma, overexpression of stromelysin-3 which was associated strictly with stromal fibroblasts adjacent to the neoplastic foci was recently reported and a correlation observed between stromelysin-3 overexpression and disease progression,[203] suggesting that this enzyme may be a useful prognostic predictor. Interestingly, a unique 10 amino acid insert within a Golgi-recognition motif which is involved in intracellular activation of the enzyme has recently been identified, suggesting that the processing of stromelysin-3 is uniquely regulated.[204]

The identification of stromelysin-3 has provided yet another important line of evidence for the critical role that the stroma can play in the regulation of tumor cell invasion. Recent co-culture studies have shown that the conditioned medium of human breast carcinoma cells contains soluble factors which can upregulate the expression of several metalloproteinases including MMP-1, 2 and 3 in human dermal fibroblasts.[205] This supports the notion that the interaction of tumor cells with the stroma is reciprocal with each component able to amplify the invasive capability of the other and contribute to tissue destruction.[194,202]

Matrilysin

In its substrate specificity and regulation, this recently identified member of the metalloproteinase family is similar to the stromelysins, however within the malignant lesions, its expression appears to be more restricted to neoplastic cells. The evidence is most extensive for carcinomas of the gastrointestinal tract particularly colon carcinoma. High expression of matrilysin, but none of stromelysins-1 and 2, was found in a high proportion of colon and gastric cancer biopsies[191] and could also be detected in a significant proportion of colonic adenomas, suggesting that upregulation of this enzyme is an early event in the progression of this disease.[201] Furthermore, the invasive and metastatic potentials of cells derived from different colon carcinoma cell lines correlated with the level of matrilysin expression and transfection of the matrilysin gene into one of these lines increased the ability of the cells to form intracecal tumors.[206,207] Conversely, suppression of matrilysin expression by retinoids or by antisense DNA caused a reduction of their invasive potential in vitro. Recently, matrilysin was also implicated in the pathogenesis of other malignancies such as prostatic and ovarian carcinomas.[208,209] Similarly to the stromelysins, this enzyme was also implicated in the activation of other metalloproteinases such as pro MMP-1.[210] Matrilysin is also expressed in monocyte precursors[211] and in proliferating human endometrial cells during the menstrual cycle, suggesting that it plays a role in processes involving tissue remodeling and cellular proliferation and migration.[212]

The identification of the metalloproteinases as major mediators of tissue degradation during invasion and metastasis has led to an effort to utilize them as prognostic markers or as targets of anti-cancer therapy. Increased expression of MMP-2 was correlated in several studies with increased incidence of local recurrence following surgery, with increased frequency of lymph node metastasis and/or with reduced survival time in patients diagnosed with diverse cancers including breast, gastric and esophageal carcinomas and squamous carcinomas of the head and neck or the oral cavity.[13] An elevation in plasma levels of MMP-2 which correlated with tumor burden was reported in lung cancer patients while an elevation in plasma levels of MMP-9 was seen in breast and colon cancer patients.[13] However, since these enzymes also play a role in diverse physiological processes which involve cell migration and tissue remodeling such as wound healing, a change in plasma levels of a particular MMP is unlikely to be a reliable independent indicator.

The extensive experimental evidence for the ability of natural and synthetic metalloproteinase inhibitors to block invasion in vitro and the increasing understanding of the chemical and molecular basis for MMP action and regulation have given impetus to a search for potent anti-metastatic inhibitors. These inhibitors can be divided into four major categories on the basis of the underlying strategy: (a) recombinant TIMP molecules or their fragments, (b) synthetic molecules which inhibit pro-enzyme activation, (c) synthetic molecules which mimic and compete with the substrate and which include a Zn^{++} chelating moiety and (d) inhibitors which interfere with signal transduction pathways involved in maintaining a constitutively high expression of the MMPs in the cancerous cells.[213] Inhibitors developed through all of these strategies have shown promise in in vitro invasion assays and were effective in preclinical metastasis studies and some have recently been approved for phase I clinical trials.[213,214] Because metalloproteinase inhibitors can also interfere with the process of angiogenesis and can have growth modulating effects on tumor cells,[215] they promise to provide a potent and relatively specific (non-toxic) class of anti cancer drugs.[213]

Role of Serine Proteinases

The Urokinase/Urokinase Receptor System of Focal Matrix Proteolysis

The urokinase plasminogen activator and its receptor also constitute an important component of the cell migration and invasion apparatus. They play a role in diverse physiological and pathological processes involving cellular migration and homing during embryogenesis, inflammation, angiogenesis, wound healing, and tumor cell invasion and metastasis.[216] Their upregulated expression was shown to be associated with increased aggressiveness of diverse tumor types and they were identified as markers of progression in various tumors including melanoma and carcinomas of the breast and prostate.[25,217-219]

Urokinase converts the enzymatically inactive glycoprotein plasminogen, found in plasma and extracellular fluids to plasmin. This broad-spectrum serine protease can mediate invasion directly by degrading ECM proteins and can also contribute to the proteolytic activity of other enzymes such as metalloproteinases, by catalyzing their conversion from latent zymogens to active enzymes.[25] Moreover, plasmin can convert the precursors of growth factors such as TGFβ to their biologically active form[220] thereby contributing to the upregulation of other proteinases such as MMP-2 and downregulation of TIMP-2 (see chapter 9, and Fig. 9.2). Tissue type plasminogen activator tPA can also catalyze the conversion of plasminogen to plasmin but its role in invasion of malignant cells appears to be more limited[24] and it will therefore not be discussed here. The reader is referred to several comprehensive reviews for more information on this enzyme.[150]

The uPA is also synthesized and secreted in an inactive single-chain precursor form, the pro-uPA.[221] In this form, uPA binds with high affinity through its amino-terminal domain, to a 55-60 kDa GPI-linked membrane receptor (uPAR) and is then converted to a two-chain active enzyme by plasmin and possibly other proteases, such as trypsin and kallikrein.[222] Plasminogen, the uPA substrate, also binds to a cell surface receptor and its conversion to plasmin occurs at the cell surface. In its bound form plasminogen is protected from the action of the circulating inhibitor α_2-antiplasmin. A third component of this cell surface proteolytic system are the PA inhibitors PAI-1, PAI-2, protease nexin 1 and the less well characterized PAI-3.[223,224] Of these inhibitors, PAI-1 is known to be an extracellular matrix protein associated with vitronectin.[225] uPAR and uPA have been localized to cell-cell and cell-substratum (focal) contact sites and plasminogen has been shown to bind to ECM proteins.[216] Thus all the components of this proteolytic apparatus are concentrated at focal contact sites where they can coordinate integrin-mediated cell adhesion with cell migration and may tilt the balance in favor of one cellular function or the other. PAI-1 can stabilize vitronectin-mediated cellular adhesion[95] thereby acting in cooperation with vitronectin receptors such as $\alpha_v\beta_3$. Vitronectin and other ECM proteins may then be degraded by plasmin resulting in cell detachment and migration.[226] Moreover, receptor bound uPA is not internalized nor processed. However, once it binds to PAI-1 via its carboxyl-terminal catalytic domain, the complex is internalized by utilizing the low-density lipoprotein (LDL) receptor-related protein (LRP), also known as the α_2 macroglobulin receptor. The LRP consists of an extracellular 515 kDa subunit containing the ligand binding region and an 85 kDa subunit composed of an extracellular, a transmembrane, and a cytoplasmic domain, the latter containing the endocytosis signal sequence, NPXY.[227-229] Following PAI-1 binding, the uPAR:uPA:PAI triplex forms a complex with the 515 kDa chain of LRP. Subsequently, the signal sequence of the 85 kDa LRP subunit allows endocytosis of the entire complex, with the uPA:PAI-1 being targeted to the lysosomes for eventual degradation and the uPAR and LRP recycled and re-expressed at new focal adhesion sites on the cell surface.[227] Binding of uPA to ECM-associated PAI-1 can therefore trigger a process of uPAR repositioning, thereby maximizing its utility in conjunction with cell migration, re-attachment and the formation of new focal adhesion contacts. While the uPA system can act in an autocrine fashion with all the essential components synthesized by the same cell, cooperativity between different cells provides an alternative paracrine mechanism in which uPAR-expressing cells bind and utilize uPA secreted by other cells. Thus tumor cells expressing high levels of uPAR have been shown to bind uPA produced by normal adjacent stromal cells or by tumor-infiltrating macrophages.[169,230]

Under normal physiological conditions uPA-mediated proteolysis, similarly to that of the metalloproteinases, is tightly regulated at the levels of synthesis and function.[226] At the transcriptional level, phorbol esters and inflammatory mediators such as TGFβ and growth factors such as EGF and bFGF,[150] which have been implicated in regulation of integrin and metalloproteinase expression, can also modulate uPAR or uPA transcription providing another link between these different components of the cell migratory apparatus. Functionally, the enzyme is secreted in a latent form requiring activation by other modulators and, in addition, its activity can be inhibited by the endogenous inhibitors PAI-1, PAI-2, the recently identified PAI-3 and protease nexin 1.[224] Deregulated expression of any of the components of this proteolytic system namely of urokinase, its receptor or the inhibitor(s) can lead to deregulated proteolysis and has been observed in numerous malignancies.[226]

As described earlier for the metalloproteinases, the evidence for the involvement of the uPA/uPAR system in tumor cell invasion at the primary site is derived from both descriptive and functional studies. In initial reports in the 1970s and 80s, a correlation was demonstrated between the secretion of high levels of uPA by virally-transformed cells and the acquisition of other properties characteristic of the malignant phenotype such as loss of anchorage-dependence for growth, increased migration and loss of cytoskeletal organization.[24] Immunohistochemical analysis of murine tumor tissue localized the enzyme to the invading front of the tumors,[231,232] and increased enzymatic activity was detected in extracts of various human tumors which were compared to the respective benign tissues, including carcinomas of the breast, colon, lung and prostate.[24] In more recent years, with the advent of molecular biology and gene transfer techniques, the development of better controlled in vitro invasion systems and the increased availability of monoclonal antibodies specific for uPA and its receptor, additional support for the critical role of uPA in invasion has been rapidly accumulating. The role of uPA in ECM degradation in vitro, which was first demonstrated by Ossowski and Reich using the chick embryo model and anti-uPA antibodies[233] was subsequently confirmed in models of human amnion and reconstituted basement membranes using cell lines of diverse tumor types including human and murine melanoma, fibrosarcoma and ovarian carcinoma[25,234] lines. Further analysis in some of these studies revealed that uPA is engaged in a proteolytic cascade where its main function (or more accurately that of plasmin) may be to mediate the conversion of latent progelatinase to its active degradative form.[25,234] Subsequently, anti-uPA antibodies injected into mice together with syngeneic tumor cells caused a significant inhibition of lung metastases formation[235] and provided the first strong functional evidence for the involvement of this enzyme in invasion and metastasis in vivo.

In more recent studies, the matrigel invasion assay was used to confirm the role of uPA and its receptor in the migration and invasion of several human-derived malignant cell lines including, melanoma, glioblastoma and ovarian prostatic and lung carcinoma.[236-238] In two recent in vivo studies, when human squamous or prostate carcinoma cells were injected into nude mice in conjunction with either anti-uPA antibodies[239] or catalytically inactive proteins which can compete for substrate binding, respectively,[240] a significant reduction in local invasiveness was seen in one case[239] and a decrease in the number of lung metastases was seen in the other.[240] This suggests that the enzyme can play a role in invasion both at the primary site and during the latter stages of the metastatic cascade.

Recent immunohistochemical analysis of human colon carcinoma specimens with antibodies to uPAR and subsequent in situ hybridization have localized high uPAR levels to the invasive foci and showed that tumor-infiltrating host cells, particularly macrophages and neutrophils were partially responsible for the strong signal.[241] Autoradiography with ^{125}I labeled pro-uPA confirmed earlier reports that expression of this receptor was significantly augmented in breast carcinoma as compared to benign breast tissue and this was reflected in increased cell surface binding of uPA.[242] While in the latter study, high levels of uPA were localized to the plasma membrane of malignant epithelial cells, a second recent immunohistochemical analysis has shown that periductal tissue macrophages surrounding infiltrating, intraductal breast carcinoma cells had a strong positive staining with antibodies to uPAR. In 8 of 50 samples tested, the carcinoma cells also stained positively but no staining was seen on in situ carcinoma or benign epithelium, identifying uPAR as a marker of progression in breast carcinoma.[243,244] Similarly, increased expression of uPAR, PAI-1 and PAI-2 and increased binding (but not expression) of uPA were recently found to be associated with progression of melanoma from the benign nevus to advanced primary melanoma.[217] uPA in this study was detected on both melanoma and stromal cells at the invading front. An association between malignant progression and production of uPA was also noted for pancreatic[245] ovarian[246] and cervical carcinoma.[247]

Together, these studies establish uPA and its receptor as markers of progression and possible diagnostic/prognostic indicators in several human malignancies including some of the most common ones. Furthermore, they confirm that similarly to the metalloproteinases, expression of components of the uPA/uPAR system is not restricted to the malignantly transformed cells and is also upregulated on adjacent stroma and infiltrating leukocytes. This provides a paracrine source of uPA for tumor cells expressing high levels of the receptor[231] and adds another dimension to the intricate stromal-epithelial interaction at the primary malignant lesion. It is pertinent to note in this context, that stimulation of uPA synthesis was noted in cultures whose breast carcinoma cell intercellular contacts were disrupted by anti E-cadherin antibodies.[248] This links uPA-mediated proteolysis to a very early event in malignant transformation namely, the disruption of cadherin mediated cell-cell contact.

Recent gene transfer experiments using full length cDNA for uPA[249] mutant uPA[240] and uPAR[251] or antisense oligonucleotides for the receptor[251,252] lend further support to the importance of these mediators in invasion and metastasis. They also provide a rationale for the development of gene therapy strategies for effective modulation of uPA-mediated proteolysis in vivo in cancer patients.

Elastase

Neutrophil elastase is another member of the family of serine proteinases which has been implicated in tumor invasion and it adds another important element to the role of the leukocytic infiltrate in local invasion.[253] Increased levels of this enzyme were found in extracts of advanced stage breast carcinoma and high levels were found to be associated with shorter disease-free and overall survival.[254] While this activity could be due to the leukocytic infiltrate in the tumor, it has been demonstrated by in vitro studies that elastase can also be produced by the breast cancer cells.[253,255] High production of this enzyme was also found to be associated with pancreatic and colon cancer.[256,257] There are few functional studies on its role in invasion but the available data suggest that in addition to elastin degradation, it probably plays an indirect role in invasion by activating and modulating the activity of other enzymes such as pro-cathepsin B[258] or uPA.[259] It is interesting to note in this respect a recent in vitro study which demonstrated that co-culture of epidermal carcinoma cells with human polymorphonuclear cells caused a detachment of the carcinoma cells from the underlying substratum and this could be blocked by an inhibitor of neutrophil elastase.[260] This provides functional evidence that the elastase can contribute to disruption of the malignant tissue.

The Role of the Cathepsins

The cathepsins are ubiquitous lysosomal proteases expressed in all mammalian cells. Two classes of cathepsins have been implicated in tumor progression, the cysteine proteinases, particularly cathepsins B and L, and the aspartic proteinase, cathepsin D (Please see chapter 10 for an overview).

Cathepsin B (CB)

Cathepsin B, the major lysosomal cysteine proteinase is optimally active at acidic pH but unlike other cysteine proteinases retains considerable activity at neutral pH and could contribute to extra-lysosomal proteolysis. In one of the earliest studies to implicate this enzyme in malignancy, Poole and associates observed that in breast carcinoma tissue, compared to benign fibroadenomas or normal breast tissue there was an increased concentration of a cathepsin B-like enzyme with catalytic activity at neutral pH.[261] Numerous studies have since confirmed that in diverse tumor types there is a correlation between CB expression, disease progression and the invasive/metastatic phenotype. This was shown in several experimental models such as metastatic variants of the B16 and other melanoma lines,[262,263] for which similar correlations could be demonstrated at both the mRNA and protein levels, in human breast carcinoma MCF-7 variants,[264] in invasive hepatoma cells[265] and in *ras*-transformed 3T3 fibroblasts which expressed increased mRNA and functional levels of both cathepsins B and L and decreased levels of their inhibitors.[266] In c-Ha-ras-transformed rat embryo fibroblasts on the other hand, no increase was seen in expression of these cysteine proteinases but a significant reduction in the level of the endogenous inhibitor was noted.[267]

Immunohistochemical analyses, in situ hybridization and functional enzyme assays carried out on tumor tissue extracts confirmed the relevance of this enzyme to human malignancies. Increased expression of this enzyme was found to correlate with disease progression in carcinomas of the lung, breast, prostate, colon and stomach as well as in gliomas.[268-273] Furthermore, elevated levels of CB and MMP-2 were recently localized to regions of tumor invasion in colon carcinoma specimens,[274] increased CB expression could be correlated with histomorphological evidence of invasion in glioma biopsies[275] and increased CB mRNA levels were localized to invasive edges of prostate carcinoma,[271] suggesting that in these malignancies CB is functionally involved in the invasive process. Interestingly, in prostate carcinoma, increased CB expression was noted in moderately invasive (Gleason grade 5-8) but not in highly invasive (grades 9 & 10) carcinomas, linking this enzyme to the early stages of stromal invasion in the prostate.[276] Similarly, in studies of breast carcinoma tissue, increased CB was detected in the local primary neoplasms but no correlation could be observed between increased enzyme expression and lymph node metastasis.[261,277]

In several separate studies with lung carcinoma patients, the CB content of the tumors was shown to be inversely correlated with disease-free survival[278,279] and it was proposed that CB levels may be useful as an independent prognostic marker for this disease. Analysis of enzyme isoforms in squamous cell carcinoma and adenocarcinoma of the lung revealed that there was a quantitative as well as a qualitative difference between the enzyme isoforms in these malignancies as compared to normal lung tissue as increased expression of the more acidic forms was found in the tumor tissue.[268] The possible prognostic value of CB in breast carcinoma is presently unclear and appears to be the subject of some controversy.[277,280]

The functional significance of elevated CB levels in malignant tissue is currently the subject of active investigation. Typically, not only the levels but also the subcellular localization of CB is modulated in tumor cells where a redistribution of the enzymes to the plasma membrane as well as its secretion to the extracellular microenvironment have been noted.[281] This change in distribution may occur as a result of alternative splicing, giving rise to variants with altered trafficking signals as has been suggested by some studies.[282] Other studies have shown that decreased pericellular pH which could conceivably occur in vivo as a result of tissue destruction,[283] or overexpression of oncogenes, such as c-Ha-ras[284] can also induce changes in intracellular localization of CB vesicles. A recent study provided evidence that in *ras*-transfected cells protein kinase C (PKC) was involved in the release of CB into the tumor periphery.[285]

Similarly to other proteolytic enzymes involved in invasion, CB may also play a dual role in ECM degradation by mediating direct proteolysis of matrix proteins on one hand and activating latent proteinases on the other. CB can degrade isolated ECM proteins such as collagens, laminin and fibronectin[286] and was recently implicated in tumor invasion through reconstituted basement membrane when it was found that cysteine proteinase inhibitors can block the invasive process[288] (also Navab R, Mort J, Brodt P, manuscript submitted). However, since specific cysteine proteinase inhibitors are presently lacking, a conclusive identification of the cysteine proteinases involved in invasion by intact cells cannot at present be achieved. CB was also implicated in proteolytic cascades and was shown to play a role in the activation of membrane-bound pro-urokinase.[289,290] In addition, cooperation between CB and metalloproteinases such as MMP-1 in the degradation of collagen[187] was demonstrated. Pro-cathepsin B, on the other hand, may be activated by uPA and leukocyte elastase[258] resulting in enhanced total proteolysis and tissue invasion through a proteolytic activation loop. It should also be noted that CB, possibly by virtue of its involvement in intracellular protein processing can also play a role in cell division and regulate growth.[291]

Cathepsin L

Cathepsin L (CL) is a broad spectrum protease with potent elastase and collagenase activities.[292] While less is known about the expression and role of this cysteine proteinase in human malignancies, experimental data suggest that it is involved in malignant progression, often together with cathepsin B.[281] The precursor for CL was initially identified as the major excreted protein (MEP) of transformed mouse fibroblasts[293] and its expression was shown to be inducible by oncogenes, tumor promoters and growth factors.[294] CL expression was found to correlate with malignant/metastatic potential of H-ras transformed murine fibroblasts,[293] murine mammary carcinoma sublines with diverse malignant potentials,[295] metastatic variants of B16 melanoma[281] and human breast carcinoma MCF-7 sublines.[264] In transformed rat embryo cells, CL expression was elevated together with that of CB[267] and similarly to CB, its expression and secretion were found to be associated with disease progression in breast[272] and colon carcinoma tissues, where it was most elevated during the early stages of invasion.[296] Similarly to CB, it was implicated in tumor invasion through basement membrane-like barriers,[297] in a proteolytic cascade with pro-uPA[290] and in the regulation of the tumorigenic phenotype.[298] While these two major lysosomal cysteine proteinases appear to be coordinately upregulated in some tumor systems, our recent evidence suggests that they can also be differentially expressed in some tumors.[299]

Cathepsin D

Much interest has been generated in the role of Cathepsin D (CD) in malignancy following reports in the late 1980s by Rocherfort and associates identifying this enzyme as an estrogen-induced product of human breast cancer cells.[300] Subsequently, a significantly increased expression of CD mRNA was detected in human breast cancer cells compared to normal mammary cells and this was reflected in higher cytosolic concentrations and higher secreted levels of the enzyme.[300,301] Further studies established CD as a marker of progression in breast carcinoma and a significant predictor of disease recurrence and overall survival.[302,303] There is presently some controversy regarding the cellular source of CD in malignant breast tissue and the relative significance of carcinoma and stromal-cell derived CD levels as prognostic indicators. Immunohistochemical analyses localized the enzyme in both reactive stroma and carcinoma cells and some studies have suggested that the level of stromal cell CD is, in fact, the more reliable prognostic indicator.[304,305]

Recent studies have widened the range of tumors in which CD production is associated with disease progression. This association has now been demonstrated in colon[306] and prostatic[307] carcinoma as well as in melanoma lesions, where increased CD production was seen in the dysplastic nevi and in more advanced melanoma,[308] suggesting that it is an early event in progression. The functional significance of increased CD production is not entirely clear as the evidence for its involvement in ECM degradation is mainly indirect.[309] It is interesting to note in this context a recent report showing that CD secreted by prostatic carcinoma cells can degrade tumor-secreted IGF-1 binding proteins.[310] Since IGF-1 and its binding proteins can rescue the cells from apoptosis and are involved in regulation of cell growth,[88] the results suggest that CD may be involved in growth modulation during cancer progression.

The Role of Motility Factors

At the primary site, the disruption of cell-cell and cell-ECM contact and the development of a matrix-degrading capability by the malignant cells is accompanied by the acquisition of a motile phenotype. Cell motility is essential for the invasive process and is mediated by diverse factors. Several growth factors which are secreted by tumor or stromal cells such as PDGF and IGF-1 can, under the proper conditions, also induce cell migration.[311,312] Stromal HGF also known as scatter factor can induce a motile phenotype in epithelial cells.[46,98] Degradative products of the ECM such as laminin, fibronectin, collagen, thrombospondin and elastin fragments[13,97,313] and hyaluronan[314] which may accumulate at the invasive front as a result of tissue degradation can also induce cell motility. These factors can trigger random cell movement (chemokinesis) or they can induce directional motility (chemotaxis) in response to a concentration gradient.[13,97] In addition, cells can also migrate towards insoluble matrix proteins such as laminin, fibronectin, type IV collagen and thrombospondin.[13,313,315] This process known as haptotaxis appears to be mediated by receptors and signal transduction pathways distinct from those involved in chemotaxis and may be the dominant mechanism during the earlier stages of invasion before tissue destruction occurs.[13]

In addition to these multifunctional factors which may be either autocrine or paracrine in origin, several motility factors produced by tumor cells have been identified and they are known collectively as autocrine motility factors (AMF). They include a 60 kDa factor which was first isolated from serum-free conditioned medium of the human melanoma cell line A2058 and which can stimulate chemotaxis of various tumor cells,[316] and a 55 kDa factor produced by B16-F1 melanoma cells and other tumor lines.[317,318] The relationship between these different factors is not at present entirely clear. Recently, a potent 125 kDa glycoprotein with motility-stimulating activity, designated autotaxin, was isolated from the A2058 melanoma line.[319] This factor could trigger both chemotactic and chemokinetic motility at picomolar concentrations and its sequence analysis revealed homology to a phosphodiesterase expressed on the surface of activated B cells and plasma cells, suggesting that these enzymes may play a role in cell motility.[320]

Morphologically and mechanistically, tumor cell motility appears to proceed in a manner similar to leukocyte migration. In response to a chemotactic stimulus, pseudopodal extensions are formed which provide the leading edge for the cell and where receptors for the appropriate motility factors and adhesion molecules can be seen to concentrate. These protrusions direct the cell in an amoeboid-like movement towards increasing concentrations of the stimulating factor.[321,322] The molecular events underlying this process of attachment and detachment are still under active investigation.[12,321,322] They are known to require assembly and disassembly of actin filaments at the migrating edge, a process involving activation of the phosphatidylinositol pathway through the activity of pertussis toxin-sensitive G proteins.[315,323]

Summary

When this body of data is viewed collectively, a picture emerges of a complex multifaceted and multi-step invasive process driven by a cascade of interactions and counter-interactions. Each mediator and each response in this cascade are at once dependent on, and a trigger for, a second mediator and response. Thus the synthesis of proteolytic, matrix-degrading enzymes such as uPA is regulated through the concerted action of adhesion receptors (both integrins and cadherins) and growth factors and these enzymes in turn can activate other proteinases (such as metalloproteinases) and release matrix-bound growth factors which can further amplify proteinase synthesis and can serve as motility-inducing factors. In addition to, or possibly as a function of, their matrix remodeling capability, members of all the families of proteinases involved in matrix dissolution also have growth modulating effects on the tumor cells.

Three major cellular compartments appear to participate in this dynamic cascade of molecular events. They include the malignantly transformed cells and the adjacent normal tissue (e.g. the normal epithelium in the case of carcinomas), the stroma and the infiltrating host leukocytes. Each of these components can activate proliferative and migratory mechanisms in the emerging malignant cells and can in turn be activated by signals released from the tumor cells, leading to continuous amplification of the process of invasion. Downregulated expression of endogenous proteinase inhibitors in response to cytokines produced by the host cells also contributes to deregulation of the entire process (see schematic representation in Fig. 11.3).

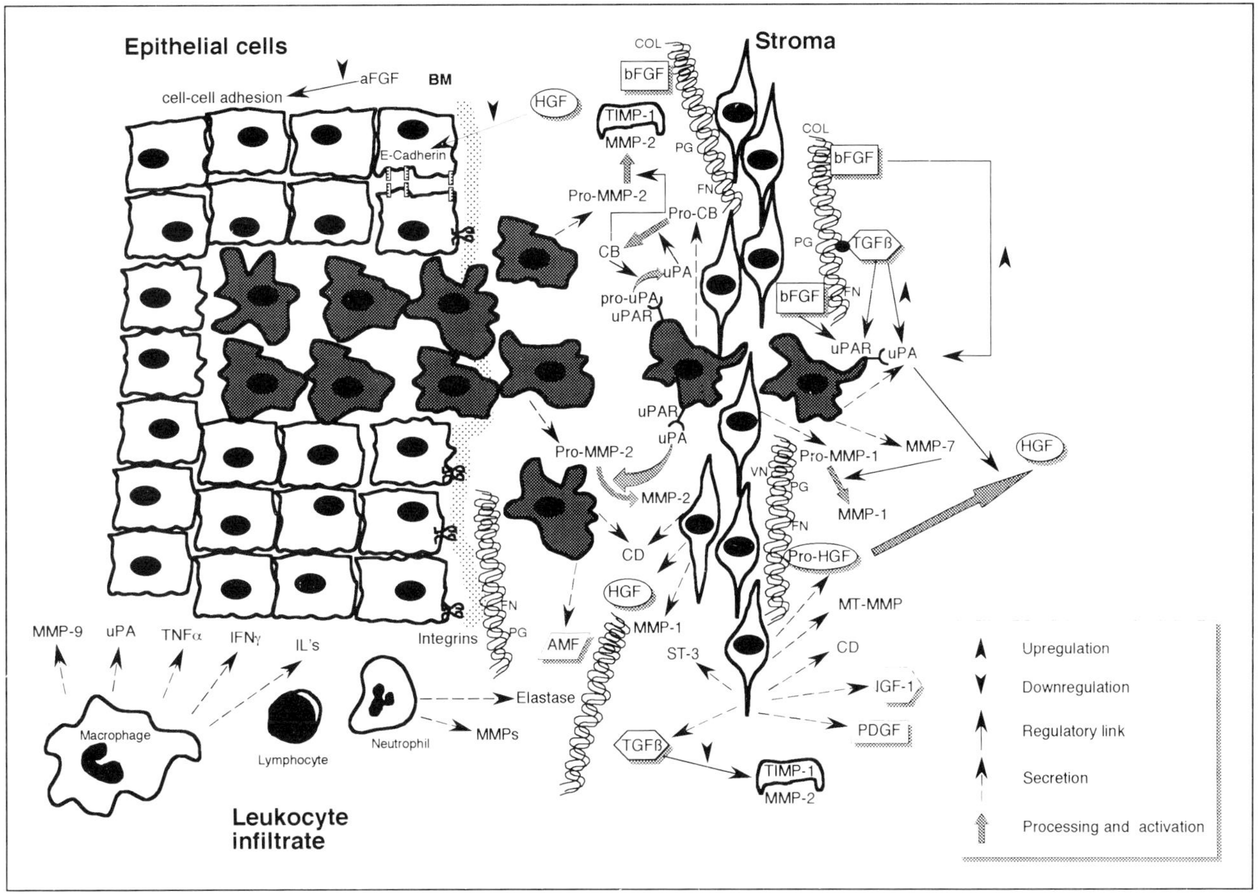

Fig. 11.3. A schematic representation of the proteolytic cascades and intercellular interactions regulating tumor cell invasion at the primary site. Shown are the cell types, molecular mediators and mechanisms of intercellular cross-talk which contribute to tumor cell motility and invasion. The abbreviations used are: aFGF; acidic fibroblast growth factor, AMF; autocrine motility factor, BM; basement membrane, CB; cathepsin B, CD; cathepsin D, HGF; hepatocyte growth factor, IFN; interferon, IGF-1; type 1 insulin-like growth factor, IL; interleukins, MMP-matrix metalloproteinase, MT; membrane type, PDGF; platelet derived growth factor, PG; proteoglycans, TIMP; tissue inhibitor of metalloproteinases, TNF; tumor necrosis factor. For additional abbreviations see legend to Fig. 11.2.

As some of the mediators which promote invasion such as uPA, type IV collagenase and their inhibitors, and several growth factors such as bFGF, TGFβ and IGF-1 also play critical roles in the process of angiogenesis, their accumulation at the invading front may be the trigger for the process of neovascularization which ultimately links the malignant cells to the vasculature and provides a channel for dissemination into distant organs. The ensuing interactions between the tumor cells and the vascular endothelium and their impact on extravasation and organ-site colonization are the subject of the concluding section of this chapter.

THE ROLE OF CELL ADHESION MECHANISMS IN TUMOR CELL INTERACTION WITH VASCULAR CELLS

The cellular mediators of adhesion and invasion which play a role in modulating the invasive/migratory phenotype of the emerging malignant cells at the primary site have also been implicated in the interactions between disseminating tumor cells, the vascular endothelium and the subendothelial matrix. As alluded to earlier, the changes which take place at the invasive front of the tumor lesion and the accumulation at this site of high concentrations of proteolytic enzymes and ECM-bound growth factors can trigger a process of neovascularization. These newly formed, tumor-induced blood vessels generally lack a well organized ECM and a protective layer of pericytes, and consequently are more permeable.[97] These, more easily penetrable vessels, provide the tumor cells with a major port of entry into the blood circulation.[324,325] Tumor cells can also utilize their ECM-degrading capability to invade the basement membranes of existing small veins and venules, or they can gain entry into lymph nodes through draining lymph vessels at the invasive edge or via the blood stream through lymphaticovenous junctions.[97,326] Once in the circulation, tumor cells may be rapidly destroyed. Their fate is ultimately determined by the consequences of their adhesive interactions with circulating blood cells such as leukocytes and platelets and with the vascular endothelial cells and the subendothelial matrix.[12,97]

Tumor Cell Interaction with Circulating Blood Cells

Interactions with Platelets

While in the circulation, tumor cells can migrate as single cells, homotypic clusters or heterotypic aggregates which are formed with blood cells such as platelets. The evidence suggests that formation of cellular aggregates provides the tumor cells with a survival advantage in the circulation, possibly by protecting them from tumoricidal host defense mechanisms, or by increasing their adhesion to the vascular endothelium and serum proteins through receptors such as the platelet integrin $\alpha_{IIb}\beta_3$.[9,16,97,326,329] Some tumors can trigger thrombin production and subsequent platelet activation. This may have several potentially beneficial effects on tumor cell survival and extravasation. The deposition of a fibrin clot around a tumor cell clump can protect tumor cells arrested in the blood vessel.[16,97] Furthermore, thrombin has recently been shown to have a cryptic RGD site which could be exposed by low concentrations of plasmin in the presence of heparan sulfate[330] and could be recognized by integrins such as $\alpha_v\beta_3$. In the presence of subendothelial heparan sulfate, thrombin can thus serve as a bridging molecule in a three way complex of tumor-platelet-endothelial cells.[330] Through the exposed RGD site it may also mediate both the anti-apoptotic effect and the activation of MMP-2 gene expression known to be triggered by cross-linking integrin $\alpha_v\beta_3$.[82] Tumor-platelet adhesion can also have a detrimental effect on tumor survival if it results in their immobilization in a hostile organ microenvironment where there is no access to growth-promoting factors or where a high concentration of growth-inhibiting cytokines is present.[97]

Interactions with Leukocytes

The interaction of tumor cells with circulating leukocytes can also have two opposing effects on their metastatic ability. While monocytes and lymphoid killer cells such as NK cells can have a tumoricidal effect on the metastasizing cells,[331] the formation of heterotypic aggregates between tumor cells and lymphocytes could potentially provide a protective, invasion-accelerating mechanism.[12,97] For example, the upregulated expression of ICAM-1 on malignant melanoma cells has been postulated to play a protective role by mediating an ICAM-1:LFA-1 interaction between the melanoma cells and circulating lymphocytes thereby linking the melanoma cells to the endothelial ICAM-1 receptor and providing a pathway for transendothelial migration.[332] Indeed on lymphoma cells, receptor LFA-1 was identified as a major mediator of cell invasion and metastasis.[15] Tumor cell interaction with polymorphonuclear cells (PMN) can also have diverse consequences. Attachment of PMN to circulating tumor cells has been observed in several animal tumor models such as B-16 melanoma and Lewis lung carcinoma[333] and could conceivably be mediated through integrin/ECM protein or LFA-1/ICAM-1 interactions. Similarly to platelets, PMNs may also serve as a bridge between the tumor cells and the vascular endothelial cells as has been shown for rat hepatocarcinoma cells.[334,335] Activation of PMNs through contact with tumor cells may trigger the release of oxygen radicals which on one hand may be toxic to the tumor cells but on the other can induce endothelial cell damage and facilitate extravasation.[336] Furthermore, PMNs can release an array of basement membrane-degrading enzymes such as MMP-2 and elastase (see above) and thereby contribute to dissolution of the subendothelial matrix and accelerate tumor cell escape from the vasculature.

Tumor cell adhesion to vascular endothelial cells

Tumor cell adhesion to the endothelial cells lining the vessel wall is a major regulatory event and a rate-limiting step in the progression of metastasis. A correlation between tumor cell adhesion to microvascular endothelial cells of specific organs in vitro and their ability to colonize the same organs in vivo has been observed in several experimental tumor models including tumors of lymphoid and epithelial origin which are selectively metastatic to the lung, liver and brain.[337] Early work in this field by Nicolson and associates was instrumental in demonstrating that once attachment occurs, it triggers endothelial cell retraction, exposing the underlying basement membrane and allowing cell migration into the subendothelial space.[1] In vivo, this migration allows the cells to extravasate into the target organ parenchyma where their growth may be promoted by local mitogenic factors.[338,339] This process involves adhesion receptors of the integrin, Ig supergene and selectin

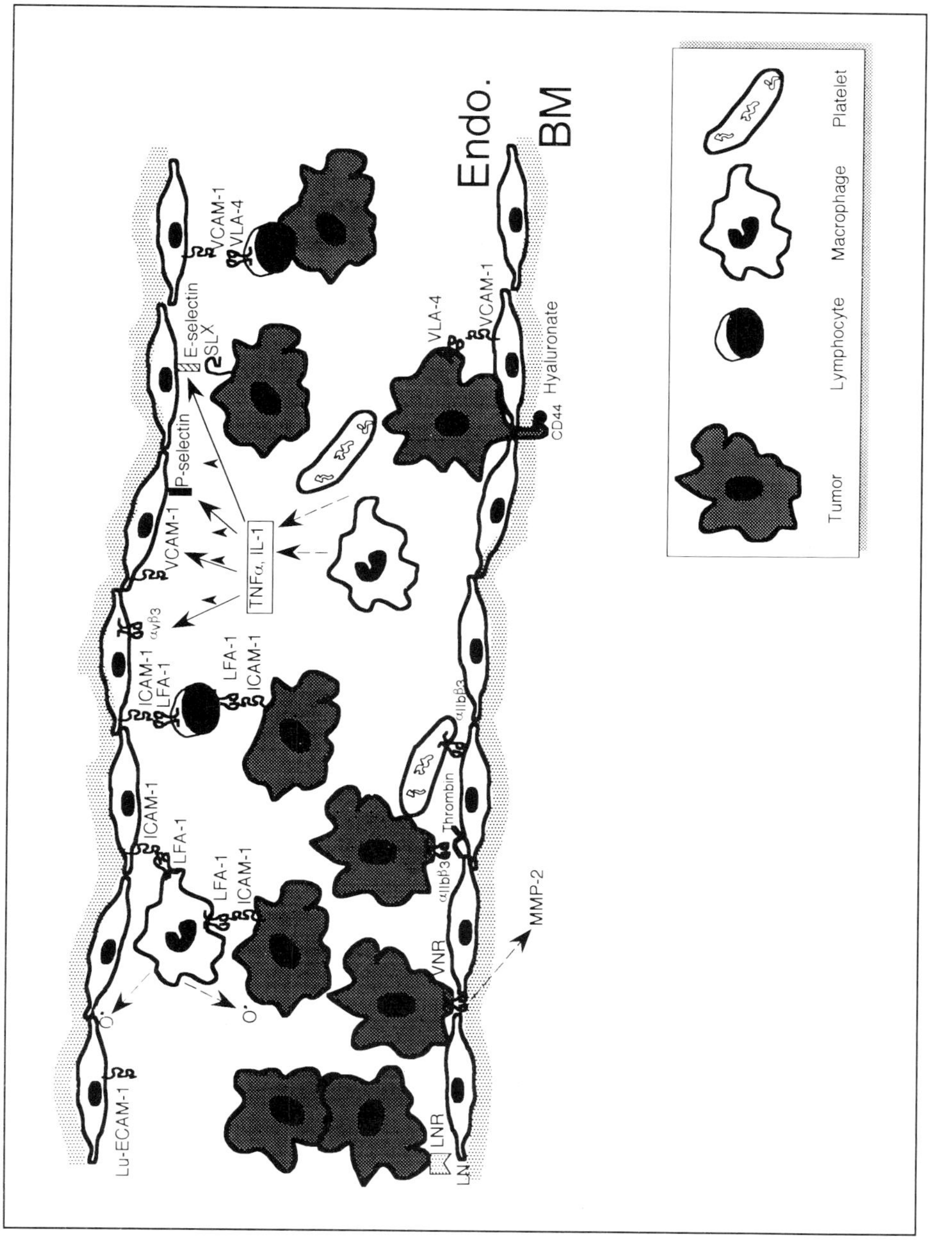

Fig. 11.4. Adhesion receptors mediating tumor cell attachment to circulating blood cells and the vascular wall. The abbreviations used are: ENDO; endothelial cells, ICAM; intercellular cell adhesion molecule, LFA; leukocyte function antigen, LNR; laminin receptor, Lu-ECAM; lung endothelial cell adhesion molecule, VCAM; vascular cell adhesion molecule, VLA; very late antigens (of T cell activation), VNR; vitronectin receptor. Other abbreviations are listed in the legends to Figs. 11.2 and 11.3.

families, often engaged in reciprocal activating cascades. It also requires the participation of basement membrane degrading proteinases (see Fig. 11.4).[1,2,12]

Based on current models of leukocyte-endothelial cell interactions which are involved in leukocyte trafficking and recruitment into sites of inflammation[340,341] (see chapters 4 and 6 for discussion), it has been postulated that tumor cell adhesion to microvascular endothelial cells also occurs in a step-wise manner. Some of the receptors involved in this interaction have been identified and initial studies suggest that they could potentially provide specific targets for therapeutic intervention.[342,343]

Much of the information on leukocyte-endothelial and tumor-endothelial cell interactions has been obtained through in vitro studies using human umbilical vein endothelial cell monolayers. These studies have shown that in response to inflammatory mediators such as TNFα and IL-1, rapid expression of several adhesion receptors is induced on the luminal side of the endothelial cell surface. These include E and P selectins, ICAM-1 and ICAM-2, VCAM-1 and several integrin receptors such as $\alpha_v\beta_3$.[341]

Some of the above molecules have been implicated in leukocyte rolling and transmigration by both in vitro and in vivo studies.[22,341] Numerous recent studies have shown that they can also mediate tumor cell adhesion to the endothelium and are likely to be involved in the process of metastasis. The evidence is particularly strong for involvement of E-selectin in tumor-endothelial interactions. E-selectin ligands, sialyl-Lewis^X and related molecules[22] have been detected on various human carcinomas. In human colorectal carcinoma for example, the expression of the sialylated Lewis^X antigen was shown to correlate with the clinical stage of the disease[344] and with metastatic potential.[345] These carbohydrates were implicated in carcinoma cell adhesion to IL-1β- or TNFα-activated human umbilical vein endothelial cells (HUVEC) through the use of specific antibodies or soluble carbohydrates.[346,347] Similar findings were described with human pancreatic carcinoma cells whose augmented adhesion to IL-1β- or TNFα-activated HUVEC could be inhibited by antibodies to E-selectin and sialyl Lewis^X.[347,348] Immunohistochemical analysis of prostatic carcinoma biopsies revealed that sialyl Lewis^X was upregulated on carcinoma cells derived from patients with aggressive metastatic disease and it was suggested that this antigen may be a valuable prognostic indicator in this disease.[349] Involvement of selectins and their ligands has also been studied in squamous cell carcinomas using immunohistochemical analysis of whole skin biopsies. These studies revealed that the dermal endothelium of squamous (SCC) and basal cell (BCC) carcinomas expressed E-selectin. However, SCC, but not BCC cells, expressed several E-selectin ligands, suggesting that the low metastatic potential of BCC may be related to the lack of expression of selectin ligands.[350] In recent in vitro studies, E-selectin was shown to mediate dynamic interactions between colon and breast carcinoma cells and vascular endothelial cells under conditions of flow, suggesting that the role of this molecule in tumor-endothelial interactions and in leukocyte trafficking may be similar.[351] Recently we have used primary cultures of murine hepatic endothelial cells to assess the role of tumor-endothelial cell adhesion in liver metastasis of the a murine Lewis lung carcinoma subline, H-59.[352] We found that TNFα-stimulated expression of E-selectin on hepatic endothelial cells, both in vitro and in vivo, and that this receptor mediated adhesion of the metastatic carcinoma cells to the TNFα-stimulated hepatic endothelial cells. Furthermore, a monoclonal antibody to murine E-selectin drastically reduced the number of experimental liver metastases when inoculated into syngeneic mice in conjunction with tumor cell injection, suggesting that E-selectin is indeed involved in liver metastasis in vivo (Brodt et al manuscript submitted). Other lines of evidence implicate cytokine-stimulated endothelial cell adhesion receptors in the organ-colonization phase of metastasis. In several studies utilizing murine or human tumor cell lines, the injection of IL-1 or TNFα into mice was shown to increase the number of experimental lung and/or liver metastases.[353-356] However, since these cytokines can activate several adhesion receptors on vascular endothelial cells, it is not always clear which receptors were engaged in the metastatic process in vivo. In vitro studies suggested that while the dominant hepatic endothelial receptor mediating liver colonization is E-selectin, other receptors such as VCAM-1 may play a more central role in lung metastasis of tumors such as melanomas.[12]

Experimental evidence suggests that members of the Ig supergene family such as VCAM-1 also play a role in tumor-endothelial cell adhesion. VCAM-1 is expressed on cytokine-activated vascular endothelial cells where it normally recognizes the counter receptors integrins $\alpha_4\beta_1$[357] (VLA-4) and probably $\alpha_4\beta_7$[358] on lymphocytes. This interaction is required for lymphocyte extravasation and diapedesis during inflammation. Upregulated expression of VLA-4 has been demonstrated in renal cell tumors,[104] in various sarcomas[359] and on melanoma cells[360-362] where expression was shown to correlate with metastatic potential.[362] VCAM-1/VLA-4 mediated adhesion of melanoma cells to HUVEC has been demonstrated by several groups[102,361] and it is thought to be the major mechanism responsible for increased lung colonization by human or murine melanoma cells following iv injections of IL-β1 or TNFα.

ICAM-1, another member of the Ig supergene family is a marker of progression in metastatic melanoma.[332] As alluded to earlier, this receptor may play a role in melanoma-leukocyte adhesion thereby facilitating melanoma attachment to, and migration through the endothelium. In addition, the importance of the LFA-1/ICAM interaction in lymphoma metastasis was recently demonstrated when experimental metastasis of murine EL4 lymphoma was inhibited by a MAb directed to the α subunit of LFA-1.[342]

Another, recently identified member of the Ig supergene family which was implicated in tumor-endothelial cell interactions is the 90 kDa lung-specific, endothelial cell adhesion molecule-1 (Lu-ECAM-1).[363,364] In vitro, this molecule selectively binds highly metastatic, lung-colonizing murine B16-F10 melanoma cells. Recent studies confirmed the relevance of this molecule to lung metastasis in vivo, as injection of a MAb directed to Lu-ECAM-1 prior to inoculation of B16-F10 cells resulted in a significant reduction in lung metastases formation.[365] Furthermore, immunohistology revealed that the staining pattern for Lu-ECAM-1 seen on endothelial cells of the pleural capillaries and sub-pleural venules in vivo paralleled the distribution of B16-F10 lung metastases,[366] suggesting that the distribution of the metastases is non-random and determined by tumor-endothelial cell interactions.

In addition to integrins $\alpha_4\beta_1$ and $\alpha_4\beta_7$, the vitronectin receptor $\alpha_v\beta_3$ may also play a role in tumor-endothelial interactions. This receptor is expressed on endothelial cells[367,368] where it is upregulated in processes such as wound healing, coagulation, and angiogenesis.[368] This integrin can mediate adhesion to several extracellular matrix proteins such as vitronectin, fibronectin, fibrinogen, thrombospondin, von Willebrand factor, ostepontin, denatured collagen, and tenascin[367] and some of these ECM proteins (i.e. vitronectin, fibronectin, and collagen) are expressed on tumor cells, providing a mechanism for enhancing tumor-endothelium adhesion during metastasis. In fact, it was recently shown that upregulation of $\alpha_v\beta_3$ expression on HUVEC by IL-1 increased adhesion of the human lung carcinoma A549 cells.[369] A recent study by Lafrenie et al demonstrated that this receptor could mediate the adhesion of various cultured human tumor cells to activated endothelial cells. These included neural crest-derived tumors such as melanomas, glioblastomas, and neuroblastomas as well as carcinomas of the bladder, breast, colon, lung, and prostate.[370] Conversely, tumor cells which express upregulated levels of integrin $\alpha_v\beta_3$ (i.e. melanomas cells) can utilize it to adhere to vitronectin in the subendothelial matrix providing another mechanism of tumor-endothelial interaction.[77] Another integrin which may act in a similar manner is $\alpha_{IIb}\beta_3$. Normally expressed on platelets, expression of this integrin on some tumor cells can be induced by phorbol esters and by the lipoxygenase metabolite of arachidonic acid 12(S) HETE (hydroxyeicosatetraenoic acid). Its expression on melanoma cells was shown to correlate with metastatic potential and the adhesion of HETE-activated melanoma cells to endothelial cells in vitro could be significantly reduced by antibodies to $\alpha_{IIb}\beta_3$ implicating it in tumor-endothelial interactions.[9] Since this integrin can mediate adhesion to several ligands including fibrinogen, fibronectin, vitronectin, thrombospondin and von Willebrand factor, it appears likely that the adhesion to endothelial cells involves bridging by one of these proteins.[9]

Integrins are also involved in attachment of tumor cells to the subendothelial matrix, an interaction which is essential for tumor cell extravasation.[12] Whereas at the primary tumor site, integrins can play a negative or positive regulatory role in tumor invasion, experimental data based on the direct injection of tumor cells intravenously (iv) into recipient mice have clearly demonstrated that integrin mediated cell-ECM interactions enhance the metastatic process once the cells are in the circulation. Synthetic RGD peptides which could block tumor cell adhesion to ECM proteins such as fibronectin, vitronectin and thrombospondin and inhibit tumor cell invasion through reconstituted basement membrane in vitro were found to abrogate lung colonization of B-16 melanoma cells which were injected through the tail vein.[371] Similar results were obtained when tumor cells were inoculated with the peptide YIGSR which blocks cell adhesion to laminin through the 67 kDa laminin receptor[372] or with anti-laminin antibodies.[373] However in other studies, a SIKVAV-containing laminin peptide was shown to promote tumor cell growth and metastases formation[374] and the injection of human or murine tumor cells together with a synthetic peptide (peptide G) which contains the laminin-binding domain of the 67 kDa receptor enhanced lung colonization.[375] These seemingly contradictory results are an indication of the complexity of tumor cell interaction with laminin, due to its multiple recognition sites and multiple roles during the progression of metastasis.[108] In this process, laminin can serve alternatively as an adhesion ligand mediating cell attachment to basement membranes, a promoter of cell locomotion and invasion and an inducer of cell growth.[108] In a recent study, Olden and Akiyama have shown that pretreatment of human breast carcinoma cells with antibodies to integrin subunits α_5 or β_1 significantly and specifically decreased their ability to form experimental lung metastasis in nude mice.[64] The same antibodies also blocked cell adhesion to, and migration on fibronectin suggesting that fibronectin plays a role in the hematogenous stage of metastasis, possibly by promoting cell extravasation into the target organ stroma.

The $\alpha_2\beta_1$ collagen/laminin receptor was identified as a positive regulator of the lung metastasizing potential of a human rhabdomyosarcoma cell line. Transfection of the α_2 subunit into the non-metastatic RD cells resulted in acquisition of a lung-colonizing potential without modulating tumorigenicity or the rate of tumor cell growth in vitro.[54] This suggests that cell adhesion to subendothelial ECM proteins may provide tumor cells with a selective advantage independently of growth modulation, possibly by promoting extravasation or by protecting the cells from circulating host immune cells, as discussed above. Integrin $\alpha_v\beta_3$, a positive regulator of melanoma growth and invasion at the primary site (see above and Fig. 11.2), may also increase melanoma cell invasion through the subendothelial ECM.[72] This receptor was also shown to mediate the adhesion of metastatic melanoma cells to vitronectin in lymph node frozen sections, suggesting that it is involved in malignant melanoma metastasis to regional draining nodes.[77,82] On tumors which express $\alpha_{IIb}\beta_3$, this integrin can also enhance adhesion to the subendothelial matrix.[9]

Another molecule which can contribute to cell adhesion to the vascular wall is CD44 because it can mediate attachment of CD44-expressing tumor cells to hyaluronan in the subendothelial matrix, thereby promoting cell locomotion.[314,376]

The ability of tumor cells to adhere to the vascular endothelium has been proposed as a major regulatory factor in the target-organ preference of metastasis seen in the clinical course of some malignant diseases and also observed in experimental models of metastasis.[1,9,12] While single specific receptors were initially thought to mediate tumor cell adhesion to the microvessel endothelium and regulate homing patterns of disseminated cells, current evidence suggests that the organ specificity is probably determined by the extent of complementation between several receptors or ligands on the endothelium and their respective counter-receptors on the tumor cells. For example, expression of E-selectin carbohydrate ligands on disseminating gastric and colonic carcinoma cells[348] may serve as the initial adhesive link to the hepatic endothelium and provide the cells with a selective advantage, but this interaction is probably only one (possibly the first) in a cascade of cell-cell adhesion events which also involve integrins and their ligands (Brodt, manuscript submitted). A recent study by Lafrenie et al clearly demonstrates this point by showing that adhesion of seven different malignant cell lines to IL-1 stimulated HUVEC could be blocked by antibodies to several different inducible adhesion receptors including E-selectin, VCAM-1 and $\alpha_v\beta_3$.[370] In vivo and in vitro data similarly suggest that adhesion of melanoma cells to the pulmonary capillary endothelium may require cooperation between several receptors such as VCAM-1, Lu-ECAM-1 and the laminin receptors. It appears therefore that the fate of the tumor cells may ultimately be determined by the cumulative effect of these adhesive interactions and their impact on the invasive/migratory capability of the tumor cells.[377]

The Role of Proteolytic Enzymes

As was discussed in detail in the preceding section, tumor cells can utilize various proteolytic enzymes to degrade extracellular matrices including basement membranes. Some of the proteolytic enzymes implicated in invasion at the primary site are also likely to play a role in degradation of the subendothelial basement membrane. This is supported by several studies in which natural and synthetic metalloproteinase inhibitors were shown to inhibit lung colonization by iv injected tumor cells. For example lung colonization by iv inoculated B16 murine melanoma cells could be inhibited by injection of rTIMP-1 into the recipient mice,[161] by ip inoculation of batimastat, a synthetic MMP inhibitor[214] or by overexpression of the inhibitor in the tumor cells.[215,378] While this was initially interpreted as evidence for the critical role of metalloproteinase-mediated invasion in lung colonization, recent evidence suggests that the role of TIMP may be more complex. In one study, overexpression of TIMP-2 in human melanoma cells was found to reduce their tumorigenicity but not their ability to metastasize spontaneously from the local site.[162] When intravital microscopy was used in another study to trace TIMP-1 transfected B16-F10 cells injected in the chick embryo chorioallantoic membrane, no change was observed in tumor cell ability to extravasate while post-extravasation growth was significantly reduced.[215] In these models the lack of effect on invasion and spontaneous lung metastasis may have been due, at least in part, to the production of other matrix-degrading proteinases such as plasmin by the melanoma cells, or to the activity of host cell-derived proteinases. More importantly, these studies show that the role of the metalloproteinases in invasion and metastasis is multi-dimensional and that matrix degradation plays an important role in cell survival and growth. This may be mediated through the release of ECM-bound growth factors or the exposure of cryptic, integrin-binding ligands which trigger signal transduction pathways essential for cell survival.[67]

In other recent studies, metastasis of human prostate carcinoma PC-3 cells to regional lymph nodes and the lung could be markedly reduced by inhibitors of the urokinase receptor,[240] and urokinase overproduction in the same cells increased skeletal metastasis following intra-cardiac injection of the tumor cells,[249] suggesting that the uPA/uPAR system plays a role in organ-colonization by these cells. Interestingly, in an earlier study by Ossowski et al using human squamous carcinoma HEp3, an antibody to uPA was found to inhibit local invasion of sc tumors but did not affect spontaneous lung metastasis.[239] These results are reminiscent of findings with MMP inhibitors and again serve to highlight the diversity of mediators which can contribute to tumor dissemination including host-cell derived proteinases. Our recent findings and those of others using the cysteine proteinase inhibitor E-64 (see above) suggest that cathepsins B and L similarly to other proteinases may play a dual role as matrix-degrading as well as growth modulating factors.[291]

Another matrix-degrading enzyme which deserves mention in this context is heparanase. Several studies by Nicolson and associates have implicated this heparan sulfate-specific endo-β-D-glucuronidase in melanoma metastasis, particularly to the brain. In these studies, nerve growth factor (NGF) was identified as a regulator of heparanase production in human and murine melanoma cells, suggesting that it may play a critical role in the high incidence of brain metastases associated with the spread of this disease.[379,380] Studies by Vlodasky and Bar-Shavit have shown that tumor as well as leukocyte-derived heparanase can degrade basement membrane heparan sulfate and in so doing contribute not only to invasion of this barrier but also to the release of heparan-sulfate-associated growth and angiogenic factors such as bFGF.[381,382] Thus heparanase can enhance organ colonization through stimulation of tumor cell migration as well as proliferation.

TUMOR CELL ADHESION TO PARENCHYMAL CELLS

The growth and motility-modulating properties of the stroma have been discussed above. Several studies indicate that in the target organ, parenchymal cells can also exert a growth-regulating effect on the disseminating tumor cells and that this cell-cell interaction may be of equal importance to that of tumor-endothelial adhesion in determining tumor potential to colonize a specific organ site.[14,15] Studies with several murine models, including our own H-59 carcinoma cells, have shown that the ability of tumor cells to metastasize to the liver correlates with adhesion to cultured hepatocytes in vitro.[14,15,383] Furthermore, we found that antibodies which blocked tumor-hepatocyte adhesion could abrogate liver colonization in vivo following intrasplenic/portal injection of the tumor cells.[14] Some of the receptors involved in this adhesion have been identified including integrin LFA-1 which was shown to mediate adhesion through ICAM-1 expressed on the hepatocytes,[15,384] integrin $\alpha_6\beta_4$ which was shown to mediate attachment of murine mammary carcinoma TA3/Ha, apparently through a cell surface determinant other than laminin[385] and integrin $\alpha_5\beta_1$ which was shown to mediate attachment through hepatocyte fibronectin.[386] This adhesion to liver parenchymal cells appears to provide the tumor cells with a growth advantage, possibly by bringing them into close proximity to hepatocyte derived, growth promoting factors.[339] Using the H-59 carcinoma model, we have recently identified IGF-1 as one liver-derived factor involved in paracrine growth stimulation of liver-metastatic cells.[387,388] Studies by other groups suggest that other factors particularly TGFα, a mediator of liver regeneration, may also be involved.[338] In addition to liver parenchymal cells, tumor cell adhesion to sinusoidal Kupffer cells was also shown to promote tumor cell potential to colonize the liver possibly by providing a bridge to the sinusoidal endothelium.[389] Together these data suggest that at least three major cellular components in the liver namely, the sinusoidal endothelium, the Kupffer cells and the hepatocytes may be involved in tumor-host cell adhesive interactions which are essential for cell survival, extravasation and growth.

CONCLUSION

A complete understanding of the cellular and molecular mechanisms underlying a process as complex and intricate as metastasis can only be achieved through a systematic analysis of the multiple host and tumor-cell derived factors and interactions mediating its progression. In recent years, this task has been facilitated through major advances in cellular, biochemical and molecular technology. These have led to the molecular cloning and biochemical characterization of many mediators involved in metastasis and to a rapidly increasing understanding of their regulation and mode of action. The recognition that many of these mediators belong to families of molecules with common structural motifs and similar functions resulted in a compartmentalization of the molecules involved in metastasis and provided insight into mechanisms of inter-molecular cross-talk. This has opened the door for development of new approaches to the design of therapeutic reagents which target specific, metastasis-promoting mediators and processes. Judging from the initial success that this intense effort has already engendered, there appears to be room for much optimism that the availability of effective and specific metastasis-retarding drugs may be a reality in the foreseeable future.

REFERENCES

1. Nicolson GL. Cancer metastasis: tumor cell and host organ properties important in metastasis to specific secondary sites. Biochim Biophys Acta 1988; 948:175-244.
2. Zetter BR. The cellular basis of site-specific tumor metastasis. New Engl J Med 1990; 322:605-12.
3. Fidler IJ. The biology of human cancer metastasis. Acta Oncol 1991; 30:668-75.
4. Liotta LA, Steeg PS, Stetler-Stevenson WG. Cancer metastasis and angiogenesis: an imbalance of positive and negative regulation. Cell 1991; 64:327-36.
5. Hart IR, Saini A. Biology of tumor metastasis. Lancet 1992; 339:1453-7.
6. McCormick BA, Zetter BR. Adhesive interactions in angiogenesis and metastasis. Pharm Ther 1992; 53:239-60.
7. Albelda SM. Role of integrins and other cell adhesion molecules in tumor progression and metastasis. Lab Invest 1993; 68:4-17.
8. Vleminckx K. Vakaet L Jr, Mareel M et al. Genetic manipulation of E-cadherin expression by epithelial tumor cells reveals an invasion suppressor role. Cell 1991; 66:107-19.
9. Honn KV, Tang DG. Adhesion molecules and tumor cell interaction with endothelium and subendothelial matrix. Cancer Metastasis Rev 1992; 11:353-75.
10. Liotta LA. The role of the extracellular matrix in tumor cell metastasis. In Vivo 1994; 8:49-58.
11. Werb Z, Tremble PM, Behrendtsen O et al. Signal transduction through fibronectin receptor induces collagenase and stromelysin gene expression. J Cell Biol 1989; 109:877-89.
12. Zetter BR. Adhesion molecules in tumor metastasis. Sem Cancer Biol 1993; 4:219-29.
13. Stetler-Stevenson WG, Aznavoorian S, Liotta LA. Tumor cell interactions with the extracellular matrix during invasion and metastasis. Ann Rev Cell Biol 1993; 9:541-73.
14. Wang J, Fallavollita L, Brodt P. Inhibition of experimental hepatic metastasis by a monoclonal antibody that blocks tumor-hepatocyte interaction. J Immunother 1994; 16:294-302.
15. Roos E, Rossien FF. Involvement of leukocyte function-associated antigen-1 (LFA-1) in the invasion of hepatocyte cultures by lymphoma and T-cell hybridoma cells. J Cell Biol 1987; 105:553-9.
16. Tohgo A, Tanaka NG, Ogawa H. Platelet aggregating activities of metastasizing tumor cells. Platelet-aggrevating activites of metastasizing tumor cells. [IV. Effects of cell surface modification on thrombin generation, platelet aggregation and subsequent lung colonization] Invasion Metastasis 1986; 6:58-68.
17. Mareel MM, Van Roy FM, Bracke ME. How and when do tumor cells metastasize? Crit Rev Oncogen 1993; 4:559-94.

18. Hynes RO. Integrins: versatility, modulation, and signaling in cell adhesion. Cell 1992; 69:11-25.
19. Juliano RL, Varner JA. Adhesion molecules in cancer: the role of integrins. Curr Opin Cell Biol 1993; 5:812-8.
20. Johnson JP. Cell adhesion molecules of the immunoglobulin supergene family and their role in malignant transformation and progression to metastatic disease. Cancer Metastasis Rev 1991; 10:11-22.
21. Takeichi M. Cadherins in cancer: implications for invasion and metastasis. Curr Opin Cell Biol 1993; 5:806-11.
22. Bevilacqua MP, Nelson RM. Selectins. J Clin Invest 1993; 91:379-87.
23. Stetler-Stevenson WG, Liotta LA, Kleiner DE Jr. Extracellular matrix 6: role of matrix metalloproteinases in tumor invasion and metastasis. FASEB J 1993; 7:1434-41.
24. Sloane BF, Moin K, Krepela E et al. Cathepsin B and its endogenous inhibitors: the role in tumor malignancy. Cancer Metastasis Rev 1990; 9:333-52.
25. Mignatti P, Rifkin DB. Biology and biochemistry of proteinases in tumor invasion. Physiol Rev 1993; 73:161-95.
26. Buxton RS, Magee AI. Structure and interactions of desmosomal and other cadherins. Sem Cell Biol 1992; 3:157-67.
27. Birchmeier W, Behrens J. Cadherin expression in carcinomas: role in the formation of cell junctions and the prevention of invasiveness. Biochim Biophys Acta 1994; 1198:11-26.
28. Gabbert H, Wagner R, Moll R et al. Tumor dedifferentiation: an important step in tumor invasion. Clin Exp Metastasis 1985; 3:257-79.
29. Takeichi M. Cadherin cell adhesion receptors as a morphogenetic regulator. Science 1991; 25:1451-5.
30. Behrens J, Mareel MM, Van Roy F et al. Dissecting tumor cell invasion: epithelial cells acquire invasive properties after the loss of uvomorulin-mediated cell-cell adhesion. J Cell Biol 1989; 108:2435-47.
31. Behrens J. The role of cell adhesion molecules in cancer invasion and metastasis. Breast Cancer and Treat 1993; 24:175-84.
32. Frixen C, Behrens J, Sachs M et al. E-cadherin mediated cell-cell adhesion prevents invasiveness of human carcinoma cell lines. J Cell Biol 1991; 111:173-85.
33. Chen WC, Obrink B. Cell-cell contacts mediated by E-cadherin uvomorulin restrict invasive behavior of L-cells. J Cell Biol 1991; 114:319-27.
34. Sommers CL, Thompson EW, Torri JA et al. Cell adhesion molecule uvomorulin expression in human breast cancer cell lines: relationship to morphology and invasive capacities. Cell Growth Diff 1991; 2:365-72.

35. Mareel MM, Behrens J, Birchmeier W et al. Down-regulation of E-cadherin expression in Madin Darby canine kidney (MDCK) cells inside tumors of nude mice. Int J Cancer 1991; 47:922-8.
36. Bussemakers MJ, van Moorselaar RJ, Giroldi LA et al. Decreased expression of E-cadherin in the progression of rat prostatic cancer. Cancer Res 1992; 52:2916-22.
37. Schipper JH, Frixen UH, Behrens J et al. E-cadherin expression in squamous cell carcinomas of head and neck: inverse correlation with tumor dedifferentation and lymph node metastasis. Cancer Res 1991; 51:6328-37.
38. Kanai Y, Oda T, Tsuda H et al. Point mutation of the E-cadherin gene in invasive lobular carcinoma of the breast. Japanese J Cancer Res 1994; 85:1035-9.
39. Piepenhagen PA, Nelson WJ. Defining E-cadherin-associated protein complexes in epithelial cells: plakoglobin, β- and γ-catenin are distinct components. J Cell Sci 1993; 104:751-62.
40. Shimoyama Y, Nagafuchi A, Fujita S et al. Cadherin dysfunction in a human cancer cell line: possible involvement of loss of α-catenin expression in reduced cell-cell adhesiveness. Cancer Res 1992; 52:5770-4.
41. Rimm DL, Sinard JH, Morrow JS. Reduced α-catenin and E-cadherin expression in breast cancer. Lab Invest 1995; 72:506-12.
42. Oyama T, Kanai Y, Ochiai A et al. A truncated β-catenin disrupts the interaction between E-cadherin and α-catenin: a cause of loss of intercellular adhesiveness in human cancer cell lines. Cancer Res 1994; 54:6282-7.
43. Behrens J, Vaket L, Friis R et al. Loss of epithelial differentiation and gain of invasiveness correlates with tyrosine phosphorylation of the E-cadherin/β-catenin complex in cells transformed with a temperature-sensitive v-src gene. J Cell Biol 1993; 120:757-66.
44. Kanai Y, Ochiai A, Shibata T et al. c-erb-2 gene product directly associates with β-catenin and plakoglobin. Biochem Biophys Res Commun 1995; 208:1067-72.
45. Boyer B, Dufour S, Thiery JP. E-cadherin expression during the acidic FGF-induced dispersion of a rat bladder carcinoma cell line. Exp Cell Res 1992; 201:347-57.
46. Weidner KM, Arakaki N, Hartmann G et al. Evidence for the identity of human scatter factor and human hepatocyte growth factor. Proc Natl Acad Sci USA 1991; 88:7001-5.
47. Weidner KM, Behrens J, Vandekerckhove J et al. Scatter factor: molecular characteristics and effect on the invasiveness of epithelial cells. J Cell Biol 1990; 111:2097-108.
48. Rubin JS, Bottaro DP, Aaronson SA. Hepatocyte growth factor/scatter factor and its receptor, the c-met proto-oncogene product. Biochim Biophys Acta 1993; 1155:357-71.
49. Aaronson SA. Growth factors and cancer. Science 1991; 254:1146-53.

50. Sorscher SM, Russack V, Graziano S et al. Immunohistochemical evaluation of E-cadherin and epidermal growth factor receptor in non-small cell lung cancer. Modern Pathol 1995; 8:450-5.
51. Shibamoto S, Hayakawa M, Takeuchi K et al. Tyrosine phosphorylation of β-catenin and plakoglobin enhanced by hepatocyte growth factor and epidermal growth factor in human carcinoma cells. Cell Adhesion Commun 1994; 1:295-305.
52. Giancotti FG, Mainiero F. Integrin-mediated adhesion and signaling in tumorigenesis. Biochim Biophys Acta 1994; 1198:47-64.
53. Pignatelli M, Vessey CJ. Adhesion molecules: novel molecular tools in tumor pathology. Human Pathology 1994; 25:849-56.
54. Chan BMC, Matsuura N, Takada Y et al. In vitro and in vivo consequences of VLA-2 expression in rhabdomyoscarcoma cells. Science 1991; 251:1600-2.
55. Giancotti FG, Ruoslahti E. Elevated levels of the $\alpha_5\beta_1$ fibronectin receptor suppress the transformed phenotype of chinese hamster ovary cells. Cell 1990; 60:849-59.
56. Dedhar S, Saulnier R. Alterations in integrin receptor expression on chemically transformed human cells: specific enhancement of laminin and collagen receptor complexes. J Cell Biol 1990; 110:481-9.
57. Plantefaber LC, Hynes RO. Changes in integrin receptors on oncogenically transformed cells. Cell 1989; 56:281-90.
58. Ruoslahti E, Giancotti FG. Integrins and tumor cell dissemination. Cancer Cells 1989; 1:119-26.
59. Bartolazzi A, Cerboni C, Nicotra MR et al. Transformation and tumor progression are frequently associated with expression of the $\alpha_3\beta_1$ heterodimer in solid tumors. Int J Cancer 1994; 58:488-91.
60. Klein CE, Steinmayer T, Kaufmann D et al. Identification of a melanoma progression antigen as integrin VLA-2. J Invest Dermatol 1991; 96:281-4.
61. Klein CE, Dressel D, Steinmayer T et al. Integrin $\alpha_2\beta_1$ is upregulated in fibroblasts and highly aggressive melanoma cells in three-dimensional collagen lattices and mediates the reorganization of collagen I fibrils. J Cell Biol 1991; 115:1427-36.
62. Gehlsen KR, Argraves WS, Pierschbacher MD et al. Inhibition of in vitro tumor cell invasion by Arg-Gly-Asp-containing synthetic peptides. J Cell Biol 1988; 106:925-30.
63. Soszka T, Knudsen KA, Beviglia L et al. Inhibition of murine melanoma cell-matrix adhesion and experimental metastasis by albolabrin, an RGD-containing peptide isolated from the venom of trimeresurus albolabris. Exp Cell Res 1991; 196:6-12.
64. Newton SA, Reeves EJ, Gralnick H et al. Inhibition of experimental metastasis of human breast carcinoma cells in athymic nude mice by anti-$\alpha_5\beta_1$ fibronectin receptor integrin antibodies. Int J Oncol 1995; 6:1063-70.

65. Chammas R, Brentani R. Integrins and metastases: an overview. Tumour Biol 1991; 12:309-20.
66. Varner JA, Fisher MH, Juliano RL. Ectopic expression of integrin $\alpha_5\beta_1$ suppresses *in vitro* growth and tumorigenicity of human colon carcinoma cells. Mol Biol Cell 1992; 3:322a (abstr.).
67. Schwartz MA. Signaling by integrins: implications for tumorigenesis. Perspectives Cancer Res 1993; 53:1503-6.
68. Akiyama SK, Larjava H, Yamada KM. Differences in the biosynthesis and localization of the fibronectin receptor in normal and transformed cultured human cells. Cancer Res 1990; 50:1601-7.
69. Zheng M, Fang H, Hakomori S. Functional role of N-glycosylation in $\alpha_5\beta_1$ integrin receptor. De-N-glycosylation induces dissociation or altered association of $\alpha_5\beta_1$ subunits and concomitant loss of fibronectin binding activity. J Biol Chem 1994; 269:12325-31.
70. Juliano RL. The role of β1 integrins in tumors. Sem Cancer Biol 1993; 4:277-83.
71. Mortarini R, Gismondi A, Santoni A et al. The role of $\alpha_5\beta_1$ integrin receptor in the proliferative response of quiescent human melanoma cells to fibronectin. Cancer Res 1992; 52:4499-506.
72. Seftor RE, Seftor EA, Stetler-Stevenson WG et al. The 72 kDa type IV collagenase is modulated via differential expression of $\alpha_v\beta_3$ and $\alpha_5\beta_1$ integrins during human melanoma cell invasion. Cancer Res 1993; 53:3411-5.
73. Albelda SM, Mette SA, Elder DE et al. Integrin distribution in malignant melanoma: association of the β3 subunit with tumor progression. Cancer Res 1990; 50:6757-64.
74. Gehlsen KR, Davis GE, Sriramarao P. Integrin expression in human melanoma cells with differing invasive and metastatic properties. Clin Exp Metastasis 1992; 10:111-20.
75. Felding-Habermann B, Mueller BM, Romerdahl CA et al. Involvement of integrin α. xv gene expression in human melanoma tumorigenicity. J Clin Invest 1992; 89:2018-22.
76. Sanders LC, Felding-Habermann B, Mueller BM et al. Role of αv integrins and vitronectin in human melanoma cell growth. Cold Spring Harbor Symposia on Quantitative Biology 1992; LVII:233-40.
77. Nip J, Shibata H, Loskutoff DJ et al. Human melanoma cells derived from lymphatic metastases use integrin $\alpha_v\beta_3$ to adhere to lymph node vitronectin. J Clin Invest 1992; 90:1406-13.
78. Montgomery AMP, Reisfeld RA, Cheresh DA. Integrin $\alpha_v\beta_3$ rescues melanoma cells from apoptosis in three-dimensional dermal collagen. Proc Natl Acad Sci USA 1994; 91:8856-60.
79. Davis GE. Affinity of integrins for damaged extracellular matrix: $\alpha_v\beta_3$ binds to denatured collagen type I through RGD sites. Biochem Biophys Res Comm 1992; 182:1025-31.

80. Seftor REB, Seftor EA, Gehlsen KR et al. Role of $\alpha_v\beta_3$ integrin in human melanoma cell invasion. Proc Natl Acad Sci USA 1992; 89:1557-61.
81. Nip, J, Rabbani SA, Shibata H et al. Coordinated expression of the vitronectin receptor and the urokinase-type plasminogen activator receptor in metastatic melanoma cells. J Clin Invest 1995; 95:2096-103.
82. Nip J, Brodt P. The role of the integrin vitronectin receptor, $\alpha_v\beta_3$ in melanoma metastasis. Cancer Metastasis Rev 1995; (in press).
83. Shih IM, Herlyn M. Role of growth factors and their receptors in the development and progression of melanoma. J Invest Derm 1993; 100:196-203.
84. Rodeck U. Growth factor independence and growth regulatory pathways in human melanoma development. Cancer Metastasis Rev 1993; 12:219-26.
85. Vuori K, Ruoslahti E. Association of insulin receptor substrate-1 with integrins. Science 1994; 266:1576-8.
86. Bartfeld NS, Pasquale EB, Geltosky JE et al. The $\alpha_v\beta_3$ integrin associates with a 190-kDa protein that is phosphorylated on tyrosine in response to platelet-derived growth factor. J Biol Chem 1993; 268:17270-6.
87. Gladson, DL, Cheresh DA. Glioblastoma expression of vitronectin and the $\alpha_v\beta_3$ integrin. Adhesion mechanism for transformed glial cells. J Clin Invest 1991; 88:1924-32.
88. Rubin R, Baserga R. Insulin-like growth factor-I: its role in cell proliferation, apoptosis, and tumorigenicity. Lab Invest 1995; 73:311-31.
89. Leavesley DI, Gerguson GD, Wayner EA et al. Requirement of the integrin β3 subunit for carcinoma cell spreading or migration on vitronectin and fibrinogen. J Cell Biol 1992; 117:1101-7.
90. Lehmann M, Rabenandrasana C, Tamura R et al. A monoclonal antibody inhibits adhesion to fibronectin and vitronectin of a colon carcinoma cell line and recognizes the integrins $\alpha_v\beta_3$, $\alpha_v\beta_5$ and $\alpha_v\beta_6$. Cancer Res 1994; 54:2102-7.
91. Lu C, Kerbel RS. Cytokines, growth factors and the loss of negative growth controls in the progression of human cutaneous malignant melanoma. Curr Opin Oncol 1994; 6:212-20.
92. Morton DL, Wong JH, Kirkwood JM et al. In: Holland JF, Frei EI, Bast RCJ et al (eds). Cancer Medicine, Lea and Febiger, Philadelphia 1993; 1793-1824.
93. Runger TM, Klein CE, Becker JC et al. The role of genetic instability, adhesion, cell motility, and immune escape mechanisms in melanoma progression. Curr Opin Oncol 1994; 6:188-96.
94. Janat MF, Argraves WS, Liau G. Regulation of vascular smooth muscle cell integrin expression by transforming growth factor β1 and by platelet-derived growth factor-BB. J Cell Physiol 1992; 151:588-95.

95. Ciambrone GJ, McKeown-Longo PJ. Plasminogen activator inhibitor type I stabilizes vitronectin-dependent adhesion in HT-1080 cells. J Cell Biol 1990; 111:2183-95.
96. Meissauer A, Kramer MD, Schirrmacher V et al. Generation of cell surface-bound plasmin by cell-associated urokinase-type or secreted tissue-type plasminogen activator: a key event in melanoma cell invasiveness *in vitro*. Exp Cell Res 1992; 199:179-90.
97. Blood CH, Zetter BR. Tumor interactions with the vasculature: angiogenesis and tumor metastasis. Biochim Biophys Acta 1990; 1032:89-118.
98. Pusztai L, Lewis CE, Lorenzen J et al. Growth factors: regulation of normal and neoplastic growth. J Pathol 1993; 169:191-201.
99. Brooks PC, Montgomery AMP, Rosenfeld M et al. Integrin $\alpha_v\beta_3$ antagonists promote tumor regression by inducing apoptosis of angiogenic blood vessels. Cell 1994; 79:1157-64.
100. Boukerche H, Benchaibi M, Berthier-Vergnes O et al. Two human melanoma cell-line variants with enhanced *in vivo* tumor growth and metastatic capacity do not express the β3 integrin subunit. Eur J Biochem 1994; 220:485-91.
101. Schadendorf D, Heidel J, Gawlik C et al. Association with clinical outcome of expression of VLA-4 in primary cutaneous malignant melanoma as well as P-selectin and E-selectin on intratumoral vessels. J Natl Cancer Inst 1995; 87:366-71.
102. Rice GE, Bevilacqua MP. An inducible endothelial cell surface glycoprotein mediates melanoma adhesion. Science 1989; 246:1303-6.
103. Martin-Padura I, Mortarini R, Lauri D et al. Heterogeneity in human melanoma cell adhesion to cytokine activated endothelial cells correlates with VLA-4 expression. Cancer Res 1991; 51:2239-41.
104. Tomita Y, Saito T, Saito K et al. Possible significance of VLA-4 ($\alpha_4\beta_1$) for hematogenous metastasis of renal-cell cancer. Int J Cancer 1995; 60:753-8.
105. Lauri D, Martin-Padura I, Biondelli T et al. Role of β1 integrins in tumor cell adhesion to cultured human endothelial cells. Lab Invest 1991; 65:525-31.
106. Romanic AM, Madri JA. The induction of 72-kD gelatinase in T cells upon adhesion to endothelial cells is VCAM-1 dependent. J Cell Biol 1994; 125:1165-78.
107. Qian F, Vaux DL, Weissman IL. Expression of the integrin $\alpha_4\beta_1$ on melanoma cells can inhibit the invasive stage of metastasis formation. Cell 1994; 77:335-47.
108. Castronovo V. Laminin receptors and laminin-binding proteins during tumor invasion and metastasis. Invasion Metastasis 1993; 13:1-30.
109. Sonnenberg A, Calfat J, Janssen H et al. Integrin $\alpha_6\beta_4$ complex is located in hemidesmosomes, suggesting a major role in epidermal cell-basement membrane adhesion. J Cell Biol 1991; 907-17.

110. Tozeren A, Kleinman HK, Wu S et al. Integrin $\alpha_6\beta_4$ mediates dynamic interactions with laminin. J Cell Sci 1994; 1:3153-63.
111. Van Waes C, Carey TE. Overexpresssion of the A9 antigen/$\alpha_6\beta_4$ integrin in head and neck cancer. Otolaryngologic Clinics of North America 1992; 25:1117-39.
112. Tennenbaum T, Weiner AK, Belanger AJ et al. The suprabasal expression of $\alpha_6\beta_4$ integrin is associated with a high risk for malignant progression in mouse skin carcinogenesis. Cancer Res 1993; 53:4803-10.
113. Friedrichs K, Ruiz P, Franke F et al. High expression level of α_6 integrin in human breast carcinoma is correlated with reduced survival. Cancer Res 1995; 55:901-6.
114. Weinel RJ, Rosendahl A, Pinschmidt E et al. The α_6-integrin receptor in pancreatic carcinoma. Gastroent 1995; 108:523-32.
115. Sager R, Anisowicz A, Neveu M et al. Identification by differential display of α_6 integrin as a candidate tumor suppressor gene. FASEB J 1993; 7:964-70.
116. Gui GP, Wells CA, Browne PD et al. Integrin expression in primary breast cancer and its relation to axillary nodal status. Surg 1995; 117:102-8.
117. Bartolazzi A, Cerboni C, Full C et al. VLA-3 distribution in normal and neoplastic non-lymphoid human tissues. Path Res Pract 1993; 189:387-93.
118. Pignatelli M, Smith MEF, Bodmer WF. Low expression of collagen receptors in moderate and poorly differentiated colorectal adenocarcinomas. Br J Cancer 1990; 61:636-8.
119. Owens LV, Xu LH, Craven RJ et al. Overexpression of the focal adhesion kinase ($_{P}125^{Fak}$) in invasive human tumors. Cancer Res 1995; 55:2752-5.
120. Hodivala KJ, Watt FM. Evidence that cadherins play a role in the downregulation of integrin expression that occurs during keratinocyte terminal differentiation. J Cell Biol 1994; 124:589-600.
121. Karecla PI, Bowden SJ, Green SJ et al. Recognition of E-cadherin on epithelial cells by the mucosal T cell integrin α(M290)β7 ($\alpha_E\beta_7$). Eur J Immunol 1995; 25:852-6.
122. Castronovo V, Luyten F, Cioce V et al. 31 kD and 12kD galactoside binding lectins are cell surface laminin binding proteins in human cancer cells: possible relationship with the 67kD laminin receptor. J Cell Biol 1990; 111:137 (abst.).
123. Pierce M, Turley EA, Roth S. Cell surface galactosyltransferase activities. Int Rev Cytol 1980; 65:1-47.
124. Jessup JM, Thomas P. Carcinoembryonic antigen: function in metastasis by human colorectal carcinoma. Cancer Metastasis Rev 1989; 8:263-80.
125. Benchimol S, Fuks A, Jothy S et al. Carcinoembryonic antigen, a human tumor marker, functions as an intercellular adhesion molecule. Cell 1989; 57:327-34.

126. Stanners CP, DeMarte L, Rojas M et al. Opposite functions for two classes of genes of the human carcinoembryonic antigen family. Tumour Biology 1995; 16:23-31.
127. Eidelman FJ, Fuks A, DeMarte L et al. Human carcinoembryonic antigen, an intercellular adhesion molecule, blocks fusion and differentiation of rat myoblasts. J Cell Biol 1993; 123:467-75.
128. Jothy S, Yuan S-Y, Shirota K. Transcription of carcinoembryonic antigen in normal colon and colon carcinoma. In situ hybridization study and implication for new in vivo functional model. Am J Pathol 1993; 143:250-7.
129. Kuroki M, Murakami M, Wakisaka M et al. Epitopes predominantly retained on the carcinoembryonic antigen molecules in plasma of patients with malignant tumors but not on those in plasma of normal individuals. Japanese J Cancer Res 1992; 83:505-14.
130. Nigam AK, Wrba F, Durbin H et al. CEA is a modulator of adhesion in colorectal tumour cells. Gut 1993; 34(suppl.#4):S67.
131. Ohannesian DW, Thomas P, Jessup M et al. Carcinoembryonic antigen and other glycoconjugates act as ligands for galectin-3 in human colon carcinoma cells. Cancer Res 1995; 55:2191-9.
132. Stoolman LM. Adhesion molecules controlling lymphocyte migration. Cell 1989; 56:907-910.
133. Gunthert U, Hofmann M, Rudy W et al. A new variant of the glycoprotein CD44 confers metastatic potential to rat carcinoma cells. Cells 1991; 65:13-24.
134. Gardner MJ, Jones LMH, Catterall JB et al. Expression of cell adhesion molecules on ovarian tumour cell lines and mesothelial cells, in relation to ovarian cancer metastasis. Cancer Letters 1995; 91:229-34.
135. Kaufmann M, Heider KH, Sinn HP et al. CD44 variant exon epitopes in primary breast cancer and length of survival. Lancet 1995; 345:615-9.
136. Ichikawa W. Positive relationship between expression of CD44 and hepatic metastases in colorectal cancer. Pathobiol 1994; 62:172-9.
137. Wielenga VJ, Heider KH, Offerhaus GJ et al. Expression of CD44 variant proteins in human colorectal cancer is related to tumor progression. Cancer Res 1993; 53:4754-6.
138. Kainz C, Kohlberger P, Sliutz G et al. Splice variants of CD44 in human cervical cancer stage IB to IIB. Gynecologic Oncol 1995; 57:383-7.
139. Seiter S, Arch R, Reber S et al. Prevention of tumor metastasis formation by anti-variant CD44. J Exp Med 1993; 177:443-55.
140. Ponta H, Sleeman J, Herrlich P. Tumor metastasis formation: cell-surface proteins confer metastasis-promoting or -suppressing properties. Biochim Biophys Acta 1994; 1198:1-10.
141. Herrlich P, Zoller M, Pals ST et al. CD44 splice variants: metastases meet lymphocytes. Immunol Today 1993; 14:395-9.

142. Wirth K, Arch R, Somasundaram C et al. Expression of CD44 isoforms carrying metastasis-associated sequences in newborn and adult rats. Eur J Cancer 1993; 29A:1172-7.
143. Arch T, Wirth K, Hofmann M et al. Participation in normal immune response of a metastasis-inducing splice variant of CD44. Science 1992; 257:682-5.
144. Aruffo A, Stamenkovic I, Melnick M et al. CD44 is the principal cell surface receptor for hyaluronate. Cell 1990; 61:1303-13.
145. St John T, Meyer J, Ederza R et al. Expression of CD44 confers a new adhesive phenotype on transfected cells. Cell 1990; 60:45-52.
146. Stamenkovic I, Aruffo A, Amiot M et al. The hematopoetic and epithelial forms of CD44 are distinct polypeptides with different adhesion potentials for hyaluronate-bearing cells. EMBO J 1991; 19:343-8.
147. Bartolazzi A, Peach R, Aruffo A et al. Interaction between CD44 and hyaluronate is directly implicated in the regulation of tumor development. J Exp Med 1994; 180:53-66.
148. Thomas L, Byers HR, Vink J et al. CD44H regulates tumor cell migration on hyaluronate-coated substrate. J Cell Biol 1992; 118:971-7.
149. Aznavoorian S, Murphy AN, Stetler-Stevenson WG et al. Molecular aspects of tumor cell invasion and metastasis. Cancer 1993; 71:1368-83.
150. Twining SS. Regulation of proteolytic activity in tissues. Critical Rev Biochem Mol Biol 1994 29:315-83.
151. Jiang WG. In vitro models of cancer invasion and metastasis: recent developments. Eur J Surg Oncol 1994; 20:493-9.
152. Birkedal-Hansen H, Moore WGI, Bodden MK et al. Matrix metalloproteinases: a review. Critical Rev Oral Biol Med 1993; 4:197-250.
153. Stetler-Stevenson WG. Type IV collagenases in tumor invasion and metastasis. Cancer Metastasis Rev 1990; 9:289-303.
154. Campo E, Merino MJ, Liotta L et al. Distribution of the 72-kD type IV collagenase in nonneoplastic and neoplastic thyroid tissue. Human Pathol 1992; 23:1395-401.
155. Jeziorska M, Haboubi NY, Schofield PF et al. Distribution of gelatinase B (MMP-9) and type IV collagen in colorectal carcinoma. Int J Colorectal Dis 1994; 9:141-8.
156. Muschel RI, Williams, JI, Lowy DR et al. *ras* modulation of metastatic potential depends upon oncogene activation and the type of recipient cell. Am J Pathol 1985; 121:1-8.
157. Garbisa S. Pozzatti R, Muschel R et al. Secretion of type IV collagenolytic protease and metastatic phenotype induction by transfection with c-Ha-ras but not c-Ha-ras plus Ad2-E1a. Cancer Res 1987; 47:1523-8.

158. Ura H, Bonfil RD, Reich R et al. Expression of type IV collagenase and procollagen genes and its correlation with the tumorigenic, invasive, and metastatic abilities of oncogene-transformed human bronchial epithelial cell line. J Natl Cancer Inst 1989; 81:587-94.
159. DeClerck Y, Perez A, Shimada H et al. Inhibition of invasion and metastasis in cells transfected with an inhibitor of metalloproteinases. Cancer Res 1992; 52:701-8.
160. Khokha R, Zimmer MJ, Graham CH et al. Suppression of invasion by inducible expression of tissue inhibitor of metalloproteinase-1 (TIMP-1) in B16-F10 melanoma cells. J Natl Cancer Inst 1992; 84:1017-22.
161. Schultz RM, Silberman S, Persky B et al. Inhibition by recombinant tissue inhibitor of metalloproteinases of human amnion invasion and lung colonization by murine B16 F10 melanoma cells. Cancer Res 1988; 48:5539-45.
162. Montgomery AM, Mueller BM, Reisfeld RA et al. Effect of tissue inhibitor of the matrix metalloproteinases-2 expression on the growth and spontaneous metastasis of a human melanoma cell line. Cancer Res 1994; 54:5467-73.
163. Poulsom R, Pignatelli M, Stetler-Stevenson WG et al. Stromal expression of 72 kDa type IV collagenase (MMP-2) and TIMP-2 mRNAs in colorectal neoplasia. Am J Pathol 1992; 141:389-96.
164. Gallegos NC, Smales C, Savage FL et al. The distribution of matrix metalloproteinases and tissue inhibitor of metalloproteinases in colorectal cancer. Surg Oncol 1995; 4:21-9.
165. Stetler-Stevenson WG, Brown PD, Onisto M et al. Tissue inhibitor of metalloproteinases-2 (TIMP-2) mRNA expression in tumor cell lines and human tumor tissues. J Biol Chem 1990; 265:13933-8.
166. Kato Y, Nakayama Y, Umeda M et al. Induction of 103- kDa gelatinase/type IV collagenase by acidic culture conditions in mouse metastatic melanoma cell lines. J Biol Chem 1992; 267:11424-30.
167. Murphy G, Docherty AJ. The matrix metalloproteinases and their inhibitors. Am J Respir Cell Mol Biol 1992; 7:120-5.
168. Kleiner DE Jr, Stetler-Stevenson, WG. Structural biochemistry and activation of matrix metalloproteases. Curr Opin Cell Biol 1993; 5:891-7.
169. Grondahl-Hansen J, Ralfkiaer E, Kirkeby LT et al. Localization of urokinase-type plasminogen activator in stromal cells in adenocarcinomas of the colon in humans. Am J Path 1991; 138:111-7.
170. Reddy VY, Zhang QY, Weiss SJ. Pericellular mobilization of the tissue-destructive cysteine proteinases, cathepsins B, L, and S, by human monocyte-derived macrophages. Proc Natl Acad Sci USA 1995; 92:3849-53.
171. Sato H, Takino T, Okada Y et al. A matrix metalloproteinase expressed on the surface of invasive tumour cells. Nature 1994; 370:61-5.

172. Okada A, Bellocq JP, Rouyer N et al. Membrane-type matrix metalloproteinase (MT-MMP) gene is expressed in stromal cells of human colon, breast, and head and neck carcinomas. Proc Natl Acad Sci USA 1995; 92:2730-4.
173. Nakajima M, Morikawa K, Fabra A et al. Influence of organ environment on extracellular matrix degradative activity and metastasis of human colon carcinoma cells. J Natl Cancer Inst 1990; 82:1890-8.
174. Waters DJ, Janovitz EB, Chan TCK. Spontaneous metastasis of PC-3 cells in athymic mice after implantation in orthotopic or ectopic microenvironments. Prostate 1995; 26:227-34.
175. Manzotti C, Audisio RA, Pratesi G. Importance of orthotopic implantation for human tumors as model systems: relevance to metastasis and invasion. Clin Exp Metastasis 1993; 11:5-14.
176. Woessner JF Jr. Matrix metalloproteinases and their inhibitors in connective tissue remodeling. FASEB J 1991; 5:2145-54.
177. Tarin D, Hoyt, BJ, Evans DJ. Correlation of collagenase secretion with metastatic-colonization potential in naturally occurring murine mammary tumors. Br J Cancer 1982; 46:266-78.
178. Whitelock JM, O'Grady RL, Gibbins JR. Interstitial collagenase (matrix metalloproteinase 1) associated with the plasma membrane of both neoplastic and nonneoplastic cells. Invasion Metastasis 1991; 11:139-48.
179. Zucker S, Wieman JM, Lysik RM et al. Metastatic mouse melanoma cells release collagen-gelatin degrading metalloproteinases as components of shed membrane vesicles. Biochem Biophys Acta 1987; 924:225-37.
180. Zucker S, Wieman JM, Lysik RM et al. Enrichment of collagen and gelatin degrading activities in the plasma membranes of human cancer cells. Cancer Res 1987; 47:1608-14.
180a. Hewitt RE, Leach IH, Powe DG et al. Distribution of collagenase and tissue inhibitor of metalloproteinases (TIMP) in colorectal tumours. Int J Cancer 1991; 49:666-72.
181. Urbanski SJ, Edwards DR, Maitland A et al. Expression of metalloproteinases and their inhibitors in primary pulmonary carcinomas. Br J Cancer 1992; 66:1188-94.
182. Gray St, Wilkins RJ, Yun K. Interstitial collagenase gene expression in oral squamous cell carcinoma. Am J Pathol 1992; 141:301-6.
183. Ute MM, Lane B, Zucker S et al. Localization of collagenase at the basal plasma membrane of a human pancreatic carcinoma cell line. Cancer Res 1990; 50:6995-7002.
184. Ossowski L. Invasion of connective tissue by human carcinoma cell lines: requirement for urokinase, urokinase receptor, and interstitial collagenase. Cancer Res 1992; 52:6754-60.
185. Crabbe T, O'Connell JP, Smith BJ et al. Reciprocated matrix metalloproteinase activation: a process performed by interstitial collagenase and progelatinase A. Biochem 1994; 33:14419-25.

186. He C, Wilhelm SM, Pentland AP et al. Tissue cooperation in a proteolytic cascade activating human interstitial collagenase. Proc Natl Acad Sci USA 1989; 86:2632-6.
187. Sires UI, Schmid TM, Fliszar CJ et al. Complete degradation of type X collagen requires the combined action of interstitial collagenase and osteoclast-derived cathepsin B. J Clin Invest 1995; 95:2089-95.
188. Wilhelm SM, Collier IE, Kronberger A et al. Human skin fibroblast stromelysin: structure, glycosylation, substrate specificity, and differential expression in normal and tumorigenic cells. Proc Natl Acad Sci USA 1987; 84:6725-9.
189. Nicholson R, Murphy G, Breathnach R. Human and rat malignant-tumor-associated mRNAs encode stromelysin-like metalloproteinases. Biochem 1989; 28:5195-203.
190. Polette M, Clavel C, Muller D et al. Detection of mRNAs encoding collagenase I and stromelysin 2 in carcinomas of the head and neck by in situ hybridization. Invasion Metastasis 1991; 11:76-83.
191. McDonnell S, Nature M, Coffey RJ Jr et al. Expression and localization of the matrix metalloproteinase pump-1 (MMP-7) in human gastric and colon carcinomas. Mol Carcinogen 1991; 4:527-33.
192. Sreenath T, Matrisian LM, Stetler-Stevenson WG et al. Expression of matrix metalloproteinase genes in transformed rat cell lines of high and low metastatic potential. Cancer Res 1992; 52:4942-7.
193. Wright JH, McDonnell S, Portella G et al. A switch from stromal to tumor cell expression of stromelysin-1 mRNA associated with the conversion of squamous to spindle carcinomas during mouse skin tumor progression. Mol Carcinogenesis 1994; 10:207-15.
194. Matrisian LM. The matrix-degrading metalloproteinases. Bioessays 1992; 14:455-63.
195. Murphy G, Cockett MI, Stephens PE et al. Stromelysin is an activator of procollagenase. Biochem J 1987; 248:265-8.
196. Unemori EN, Bair MJ, Bauer EA et al. Stromelysin expression regulates collagenase activation in human fibroblasts. J Biol Chem 1991; 266:23477-82.
197. Koklitis PA, Murphy G, Sutton C et al. Purification of recombinant human prostromelysin. Biochem J 1991; 276:217-221.
198. Basset P, Bellocq JP, Wolf C et al. A novel metalloproteinase gene specifically expressed in stromal cells of breast carcinomas. Nature 1990; 348:699-704.
199. Muller D, Wolf C, Abecassis J et al. Increased stromelysin 3 gene expression is associated with increased local invasiveness in head and neck squamous cell carcinomas. Cancer Res 1993; 53:165-9.
200. Majmudar G, Nelson BR, Jensen TC et al. Increased expression of stromelysin-3 in basal cell carcinomas. Mol Carcinogenesis 1994; 9:17-23.
201. Newell KJ, Witty JP, Rodgers WH et al. Expression and localization of matrix-degrading metalloproteinases during colorectal tumorigenesis. Mol Carcinogenesis 1994; 10:199-206.

202. Basset P, Rouyer N, Wolf C et al. Stromelysin-3—a paradigm of extracellular proteinase expressed in stromal cells of human carcinomas. Cell Biol Int 1995; 19:242.
203. Porte H, Chastre E, Prevot S et al. Neoplastic progression of human colorectal cancer is associated with overexpression of the stromelysin-3 and BM-40/Sparc genes. Int J Cancer 1995; 64:70-5.
204. Pei DQ, Weiss SJ. Furin-dependent intracellular activation of the human stromelysin-3 zymogen. Nature 1995; 375:244-7.
205. Ito A, Nakajima S, Sasaguri Y et al. Co-culture of human breast adenocarcinoma MCF-7 cells and human dermal fibroblasts enhances the production of matrix metalloproteinases 1, 2 and 3 in fibroblasts. Br J Cancer 1995; 71:1039-45.
206. Witty JP, McDonnell S, Newell KJ et al. Modulation of matrilysin levels in colon carcinoma cell lines affects tumorigenicity in vivo. Cancer Res 1994; 54:4805-12.
207. Newell KJ, Witty JP, Rodgers WH et al. Expression and localization of matrix-degrading metalloproteinases during colorectal tumorigenesis. Mol Carcinogenesis 1994; 10:199-206.
208. Knox JD, Mack CF, Powell WC et al. Prostate tumor cell invasion: a comparison of orthotopic and ectopic models. Invasion Metastasis 1993; 13:325-31.
209. Matrisian LM. Matrix metalloproteinase gene expression. Ann NY Acad Sci 1994; 732:42-50.
210. Imai K, Yokohama Y, Nakanishi I et al. Matrix metalloproteinase 7 (matrilysin) from human rectal carcinoma cells. Activation of the precursor, interaction with other matrix metalloproteinases and enzymic properties. J Biol Chem 1995; 270:6691-7.
211. Busiek DF, Ross FP, McDonnel S et al. The matrix metalloprotease matrilysin (PUMP) is expressed in developing human mononuclear phagocytes. J Biol Chem 1992; 267:9087-92.
212. Rodgers WH, Osteen KG, Matrisian LM et al. Expression and localization of matrilysin, a matrix metalloproteinase, in human endometrium during the reproductive cycle. Am J Obstet Gynecol 1993; 168:253-60.
213. Kohn EC, Liotta LA. Molecular insights into cancer invasion: strategies for prevention and intervention. Cancer Res 1995; 55:1856-62.
214. Wang X, Fu X, Brown PD et al. Matrix metalloproteinase inhibitor BB-94 (Batimastat) inhibits human colon tumor growth and spread in a patient-like orthotopic model in nude mice. Cancer Res 1994; 54:4726-8.
215. Koop S, Khokha R, Schmidt EE et al. Overexpression of metalloproteinase inhibitor in B16F10 cells does not affect extravasation but reduces tumor growth. Cancer Res 1994; 54:4791-7.
216. Vassalli J-D, Sappini A-P, Belin D. The plasminogen activator/plasmin system. J Clin Invest 1991; 88:1067-72.

217. de Vries TJ, Quax PH, Denijn M et al. Plasminogen activators, their inhibitors and urokinase receptor emerge in late stages of melanocytic tumor progression. Am J Path 1994; 144:70-81.
218. Delbaldo C, Masouye I, Surat J-H et al. Plasminogen activation in melanocytic neoplasia. Cancer Res 1994; 54:4547-52.
219. Kwaan HC. The plasminogen-plasmin system in malignancy. Cancer Metastasis Rev 1992; 11:291-311.
220. Flaumenhaft R, Abe M, Mignatti P et al. Basic fibroblast growth factor-induced activation of latent transforming growth factor β in endothelial cells: regulation of plasminogen activator activity. J Cell Biol 1992; 118:901-9.
221. Skriver L, Nielsen LS, Stephens R et al. Plasminogen activator released as inactive proenzyme from sarcoma virus transformed murine cells. Eur J Biochem 1982; 124:409-14.
222. Ellis V, Dan K. Plasminogen activation by receptor-bound urokinase. Sem Thromb Hemostasis 1991; 17:194-200.
223. Heeb MJ, Espana F, Geiger M et al. Immunological identity of heparin-dependent plasma and urinary protein C inhibitor and plasminogen activator inhibitor-3. J Biol Chem 1987; 262:15813-6.
224. Kruithof EK. Plasminogen activator inhibitors—a review. Enzyme 1988; 40:113-21.
225. Seiffert D, Mimuro J, Schleef RR et al. Interactions between type I plasminogen activator inhibitor, extracellular matrix and vitronectin. Cell Diff Dev 1990; 32:287-92.
226. Blasi F. Urokinase and urokinase receptor: a paracrine/autocrine system regulating cell migration and invasiveness. Bioessays 1993; 15:105-11.
227. Herz J, Couthier DE, Hammer RE. LDL receptor-related protein internalizes and degrades uPA-PA-1 complexes and is essential for embryo implantation. Cell 1992; 71:411-21.
228. Olson D, Pollanen J, Hoyer-Hansen G et al. Internalization of the urokinase-plasminogen activator inhibitor type-1 complex is mediated by the urokinase receptor. J Biol Chem 1992; 267:9129-33.
229. Bu G, Warshawsky I, Schwartz AL. Cellular receptors for plasminogen activators. Blood 1994; 83:3427-36.
230. Pyke C, Kirstensen P, Ralfkier E et al. Urokinase-type plasminogen activator is expressed in stromal cells and its receptor in cancer cells at invasive foci in human colon adenocarcinomas. Am J Path 1991; 138:1059-67.
231. Skriver L, Larsson LI, Kielberg V et al. Immunocytochemical localization of urokinase-type plasminogen activator in Lewis lung carcinoma. J Cell Biol 1984; 99:752-7.
232. Camiolo SM, Markus G, Englander LS et al. Plasminogen activator content and secretion in explants of neoplastic and benign human prostate tissues. Cancer Res 1984; 44:311-8.
233. Ossowski L, Reich E. Antibodies to plasminogen activator inhibit human tumor metastasis. Cell 1983; 35:611-9.

234. Reich R, Thompson EW, Iwamoto Y et al. Effects of inhibitors of plasminogen activator, serine proteinases, and collagenase IV on the invasion of basement membranes by metastatic cells. Cancer Res 1988; 48:3307-12.
235. Hearing VJ, Law LW, Corti A et al. Modulation of metastatic potential by cell surface urokinase of murine melanoma cells. Cancer Res 1988; 48:1270-8.
236. Kobayashi H, Ohi H, Shinohara H et al. Saturation of tumour cell surface receptors for urokinase-type plasminogen activator by amino-terminal fragment and subsequent effect on reconstituted basement membranes invasion. Br J Cancer 1993; 67:537-44.
237. Wilhelm O, Weidle U, Hohl S et al. Recombinant soluble urokinase receptor as a scavenger for urokinase-type plasminogen activator (uPA). Inhibition of proliferation and invasion of human ovarian cancer cells. FEBS Letters 1994; 337:131-4.
238. Liu G, Shuman MA, Cohen RL. Co-expression of urokinase, urokinase receptor and PAI-1 is necessary for optimum invasiveness of cultured lung cancer cells. Int J Cancer 1995; 60:501-6.
239. Ossowski L, Russo-Payne H, Wilson EL. Inhibition of urokinase-type plasminogen activator by antibodies: the effect on dissemination of a human tumor in the nude mouse. Cancer Res 1991; 51:274-81.
240. Crowley CW, Cohen RL, Lucas BK et al. Prevention of metastasis by inhibition of the urokinase receptor. Proc Natl Acad Sci USA 1993; 90:5021-5.
241. Pyke C, Ralfkiaer E, Ronne E et al. Immunohistochemical detection of the receptor for urokinase plasminogen activator in human colon cancer. Histopath 1994; 24:131-8.
242. Del Vecchio S, Stoppelli MP, Carriero MV et al. Human urokinase receptor concentration in malignant and benign breast tumors by in vitro quantitative autoradiography: comparison with urokinase levels. Cancer Res 1993; 53:3198-206.
243. Pyke C, Graem N, Ralfkiaer E. Receptor for urokinase is present in tumor-associated macrophages in ductal breast carcinoma. Cancer Res 1993; 53:1911-5.
244. Brunner N, Pyke C, Hansen CH et al. Urokinase plasminogen activator (uPA) and its type 1 inhibitor (PAI-1): regulators of proteolysis during cancer invasion and prognostic parameters in breast cancer. Cancer Treat Res 1994; 71:299-309.
245. Takeuchi Y, Nakao A, Harada A et al. Expression of plasminogen activators and their inhibitors in human pancreatic carcinoma: immunohistochemical study. Am J Gastroent 1993; 88:1928-33.
246. Kuhn W, Pache L, Schmalfeldt B et al. Urokinase (uPA) and PAI-1 predict survival in advance ovarian cancer patients (FIGO III) after radical surgery and platinum-based chemotherapy. Gynecol Oncol 1994; 55:401-9.

247. Kobayashi H, Fujishiro S, Terao T. Impact of urokinase-type plasminogen activator and its inhibitor type 1 on prognosis in cervical cancer of the uterus. Cancer Res 1994; 54:6539-48.
248. Frixen UH, Nagamine Y. Stimulation of urokinase-type plasminogen activator expression by blockage of E-cadherin-dependent cell-cell adhesion. Cancer Res 1993; 53:3618-23.
249. Achbarou A, Kaiser S, Tremblay G et al. Urokinase overproduction results in increased skeletal metastasis by prostate cancer cells in vivo. Cancer Res 1994; 54:2372-7.
250. Kariko K, Kuo A, Boyd D et al. Overexpression of urokinase receptor increases matrix invasion without altering cell migration in a human osteosarcoma cell line. Cancer Res 1993; 53:3109-17.
251. Kook YH, Adamski J, Zelent A et al. The effect of antisense inhibition of urokinase receptor in human squamous cell carcinoma on malignancy. EMBO J 1994; 13:3983-91.
252. Quattrone A, Fibbi G, Anichini E et al. Reversion of the invasive phenotype of transformed human fibroblasts by anti-messenger oligonucleotide inhibition of urokinase receptor gene expression. Cancer Res 1995; 55:90-5.
253. Kay G, Kennedy R, Walker B et al. The detection of serine elastase in human breast cancer. Biochem Soc Transac 1994; 22:20S.
254. Yamashita J, Ogawa M, Shirakusa T. Free-form neutrophil elastase is an independent marker predicting recurrence in primary breast cancer. J Leuk Biol 1995; 57:375-8.
255. Yamashita JI, Ogawa M, Ikei S et al. Production of immunoreactive polymorphonuclear leucocyte elastase in human breast cancer cells: possible role of ploymorphonuclear leucocyte elastase in the progression of human breast cancer. Br J Cancer 1994; 69:72-6.
256. Iwase K, Miyata M, Yamaguchi T et al. Pancreatic ductal cell carcinoma producing pancreatic elastase 1. J Surg Oncol 1993; 54:199-202.
257. Saitoh O, Sugi K, Matsuse R et al. The forms and the levels of fecal PMN-elastase in patients with colorectal diseases. Am J Gastroenterol 1995; 90:388-93.
258. Dalet-Fumeron V, Guinec N, Pagano M. In vitro activation of procathepsin B by three serine proteinases: leucocyte elastase, cathepsin G, and the urokinase-type plasminogen activator. FEBS Letters 1993; 332:251-4.
259. Schmitt M, Kanayama N, Janicke F et al. Human tumor cell urokinase-type plasminogen activator (uPA): degradation of the proenzyme form (pro-uPA) by granulocyte elastase prevents subsequent activation by plasmin. Adv Exp Med Biol 1991; 297:111-28.
260. Katayama H, Kitagawa S, Masuyama J et al. Polymorphonuclear leukocyte-induced detachment of cultured epidermal carcinoma cells from the substratum. J Invest Derm 1991; 97:949-52.

261. Poole AR, Tiltman KJ, Recklies AD et al. Differences in secretion of the proteinase cathepsin B at the edges of human breast carcinomas and fibroadenomas. Nature (London) 1978; 273:545-7.
262. Sloane BF, Dunn JF, Honn KV. Lysosomal cathepsin B: correlation with metastatic potential. Science 1981; 212:1151-3.
263. Sloane BF, Rozhin J, Johnson K et al. Cathepsin B: association with plasma membrane in metastatic tumors. Proc Nat Acad Sci USA 1986; 83:2483-7.
264. Scadden PB, Dufresne MJ. Characterization of cysteine proteases and their endogenous inhibitors in MCF-7 and adriamycin-resistant MCF-7 human breast cancer cells. Invasion Metastasis 1993; 13:301-13.
265. Moin K, Rozhin J, McKernan TB et al. Enhanced levels of cathepsin B mRNA in murine tumors. FEBS Letters 1989; 244:61-4.
266. Chambers AF, Colella R, Denhardt DT et al. Increased expression of cathepsins L and B and decreased activity of their inhibitors in metastatic, ras-transformed NIH 3Y3 cells. Mol Carcinogen 1992; 5:238-45.
267. Sloane BF, Rozhin J, Moin K et al. Cysteine endopeptidases and their inhibitors in malignant progression of rat embryo fibroblasts. Biol Chem Hoppe-Seyler 1992; 373:589-94.
268. Krepela E, Prochazka J, Mynarikova H et al. Multiple forms of cathepsin B in human lung cancer. Int J Cancer 1995; 61:44-53.
269. Matsuoka Y, Tsushima H, Koga Y et al. An inactive cathepsin B-like enzyme and cysteine proteinase inhibitors in colon cancer ascites. Neoplasma 1992; 39:107-14.
270. Hirano T, Yoshioka H. Serum cathepsin B levels, urinary excretion of cathepsin B and tissue cathepsin B content in the patients with gastric cancer. Nippon Geka Hokan—Archiv Japanische Chirurgie 1993; 62:217-21.
271. Sinha AA, Gleason DF, Deleon OF et al. Localization of a biotinylated cathepsin B oligonucleotide probe in human prostate including invasive cells and invasive edges by in situ hybridization. Anatom Rec 1993; 235:233-40.
272. Gabrijelcic D, Svetic B, Spaic D et al. Cathepsins B, H and L in human breast carcinoma. Eur J Clin Chem Clin Biochem 1992; 30:69-74.
273. Sivaparvathi M, Sawaya R, Wang SW et al. Overexpression and localization of cathepsin B during the progression of human gliomas. Clin Exp Metastasis 1995; 13:49-56.
274. Emmert-Buck MR, Roth MJ, Zhuang Z et al. Increased gelatinase A (MMP-2) and cathepsin B activity in invasive tumor regions of human colon cancer samples. Am J Pathol 1994; 145:1285-90.
275. Rempel SA, Rosenblum ML, Mikkelsen T et al. Cathepsin B expression and localization in glioma progression and invasion. Cancer Res 1994; 54:6027-31.

276. Sinha AA, Wilson MJ, Gleason DF et al. Immunohostocheimcal localization of cathepsin B in neoplastic human prostate. Prostate 1995; 26:171-8.
277. Gabrijelcic D, Svetic B, Spaic D et al. Determination of cathepsins B, H, L and kininogen in breast cancer patients. Agents Actions 1992; 38:350-7.
278. Sukoh N, Abe S, Ogura S et al. Immunohistochemical study of cathepsin B. Prognostic significance in human lung cancer. Cancer 1994; 74:46-51.
279. Ebert W, Knoch H, Werle B et al. Prognostic value of increased lung tumor tissue cathepsin B. Anticancer Res 1994; 14:895-9.
280. Castiflioni T, Merino MJ, Elsner B et al. Immunohistochemical analysis of cathepsins D, B and L in human breast cancer. Human Path 1994; 25:857-62.
281. Sloane BF, Rozhin J, Krepela E et al. The malignant phenotype and cysteine proteinases. Biomed Biochim Acta 1991; 50:549-54.
282. Sloane BF, Sameni M, Cao L et al. Alterations in processing and trafficking of cathepsin B during malignant progression. In: Katunuma N, Suzuki K, Travis J et al, eds. Biological Functions of Proteases and Inhibitors. Tokyo: Japan Scientific Societies Press 1994; 131-47.
283. Rozhin J, Sameni M, Ziegler G et al. Pericellular pH affects distribution and secretion of cathepsin B in malignant cells. Cancer Res 1994; 54:6517-25.
284. Sloane BF, Moin K, Sameni M et al. membrane association of cathepsin B can be induced by transfection of human breast epithelial cells with c-Ha-ras oncogene. J Cell Sci 1994; 107:373-84.
285. Hohn KV, Timar J, Rozhin J et al. A lipoxygenase metabolite, 12-(S)-HETE, stimulates protein kinase C-mediated release of cathepsin B from malignant cells. Exp Cell Res 1994; 214:120-30.
286. Cardozo C. Degradation of rat lung collagens by cathepsin B. J Lab Clin Med 1992; 119:169-75.
287. Buck MR, Karustis DG, Day NA et al. Degradation of extracellular-matrix proteins by human cathepsin B from normal and tumour tissues. Biochem J 1992; 282:273-8.
288. Redwood SM, Lui BC, Weiss RE et al. Abrogation of the invasion of human bladder tumor cells by using protease inhibitor(s). Cancer 1992; 69:1212-9.
289. Kobayashi H, Moniwa N, Sugimur M et al. Effects of membrane-associated cathepsin B on the activation of receptor-bound prourokinase and subsequent invasion of reconstituted basement membranes. Biochim Biophys Acta 1993; 1178:55-62.
290. Schmitt M, Janicke F, Moniwa N et al. Tumor-associated urokinase-type plasminogen activator: biological and clinical significance. Biol Chem Hoppe-Seyler 1992; 373:611-22.

291. Shoji-Kasai Y, Senshu M, Iwashita S et al. Thiol protease-specific inhibitor E-64 arrests human epidermoid carcinoma A431 cells at mitotic metaphase. Proc Natl Acad Sci USA 1988; 85:146-50.
292. Chauhan SS, Popescu NC, Ray D et al. Cloning, genomic organization, and chromosomal localization of human cathepsin L. J Biol Chem 1993; 268:1039-45.
293. Denhardt DT, Greenberg AH, Egan SE et al. Cathepsin L expression correlates closely with the metastatic potential of H-ras-transformed murine fibroblasts. Oncogene 1987; 2:55-9.
294. Troen BR, Chauhan SS, Ray D et al. Downstream sequences mediate induction of the mouse cathepsin L promoter by phorbol esters. Cell Growth Diff 1991; 2:23-31.
295. Morris VL, Tuck AB, Wilson SM et al. Tumor progression and metastasis in murine D2 hyperplastic alveolar nodule mammary tumor cell lines. Clin Exp Metastasis 1993; 11:103-12.
296. Shuja S, Sheahan K, Murnane MJ. Cysteine endopeptidase activity levels in normal human tissues, colorectal adenomas and carcinomas. Int J Cancer 1991; 49:341-6.
297. Yagel S, Warner AH, Nellans HN et al. Suppression by cathepsin L inhibitors of the invasion of amnion membranes by murine cancer cells. Cancer Res 1989; 49:3553-7.
298. Weber E, Gunther D, Laube F et al. Hybridoma cells producing antibodies to cathepsin L have greatly reduced potential for tumour growth. J Cancer Res Clin Oncol 1994; 120:564-7.
299. Brodt P, Reich R, Moroz LA et al. Differences in the repertoires of basement membrane degrading enzymes in two carcinoma sublines with distinct patterns of site-selective metastasis. Biochim Biophys Acta 1992; 1139:77-83.
300. Capony F, Morisset M, Barrett AJ et al. Phosphorylation, glycosylation and proteolytic activity of the 52kD estrogen-induced protein secreted by MCF7 cells. J Cell Biol 1987; 104:253-62.
301. Capony F, Rougeot C, Montcourrier P et al. Increased secretion, altered processing, and glycosylation of pro-cathepsin D in human mammary cancer cells. Cancer Res 1989; 49:3904-9.
302. Gion M, Mione R, Dittadi R et al. Relationship between cathepsin D and other pathological and biological parameters in 1752 patients with primary breast cancer. Eur J Cancer 1995; 31A:671-7.
303. Schwartz MK. Tissue cathepsins as tumor markers. Clin Chim Acta 1995; 237:67-78.
304. Joensuu H, Toikkanen S, Isola J. Stromal cell cathepsin D expression and long-term survival in breast cancer. Br J Cancer 1995; 71:155-9.
305. Walker RA, Denley H, Dookeran KA. Cathepsin D in breast carcinomas—the role of the stromal cell component. Oncol Reports 1994; 1:227-31.

306. Tumminello FM, Gebbia N, Pizzolanti G et al. Cathepsin D content in colorectal cancer– correlation with cathepsin D activity and other biological parameters– a preliminary report. Oncol 1995; 52:237-42.
307. Kuczyk MA, Serth J, Bokemeyer C et al. Immunohistochemical detection of cathepsin D expression in prostate cancer. Oncol Reports 1994; 1:1247-51.
308. Podhajcer OL, Bover L, Bravo AI et al. Expression of cathepsin D in primary and metastatic human melanoma and dysplastic nevi. J. Invest Dermtol 1995; 104:340-4.
309. Rochefort H. Overexpression, derouting and metastatic potential of cathepsin D in breast cancer. In: Lippman ME and Mihich E, eds. The Therapeutic Implications of the Molecular Biology of Breast Cancer. New York: John Libbey CIC 1991:267.
310. Conover CA, Perry JE, Tindall DJ. Endogenous Cathepsin D-mediated hydrolysis of insulin-like growth factor-binding proteins in cultured human prostatic carcinoma cells. J Clin Endocrin Metabol 1995; 80:987-93.
311. Kundra V, Escobedo J, Kazlauskas A et al. Regulation of chemotaxis by the platelet-derived growth factor receptor-β. Nature 1994; 367:474-6.
312. Stracke ML, Engel JD, Wilson W et al. The type 1 insulin-like growth factor receptor is a motility receptor in human melanoma cells. J Biol Chem 1989; 264:1544-9.
313. Lester BR, McCarthy JB. Tumor cell adhesion to the extracellular matrix and signal transduction mechanisms implicated in tumor cell motility, invasion and metastasis. Cancer Metastasis Rev 1992; 11:31-44.
314. Turley EA. Hyaluronan and cell locomotion. Cancer Metastasis Rev 1992; 11:21-30.
315. Aznavoorian S, Stracke ML, Krutzsch et al. Signal transduction for chemotaxis and haptotaxis by matrix macromolecules in tumor cells. J Cell Biol 1990; 110:1427-38.
316. Liotta LA. Tumor invasion and metastases-role of the extracellular matrix. Cancer Res 1986; 46:1-7.
317. Nabi IR, Watanabe H, Raz A. Autocrine motility factor and its receptor: role in cell locomotion and metastasis. Cancer Metastasis Rev 1992; 11:5-20.
318. Silletti S, Raz A. Autocrine motility factor is a growth factor. Biochem Biophys Res Commun 1993; 194:446-57.
319. Stracke ML, Krutzsch HC, Unsworth EJ et al. Identification, purification, and partial sequence analysis of autotaxin, a novel motility-stimulating protein. J Biol Chem 1992; 267:2524-9.
320. Murata J, Lee HY, Clair T et al. cDNA cloning of the human tumor motility-stimulating protein, autotaxin, reveals a homology with phosphodiesterases. J Biol Chem 1994; 269:30479-84.

321. Lester BR, McCarthy JB, Sun Z et al. G-protein involvement in matrix-mediated motility and invasion of high and low experimental metastatic B16 melanoma clones. Cancer Res 1989; 49:5940-8.
322. Condeelis J, Jones J, Segall JE. Chemotaxis of metastatic tumor cells: Clues to mechanisms from the Dictyostelium paradigm. Cancer Metastasis Rev 1992; 11:55-68.
323. Lester BR, Weinstein LS, McCarthy JB, et al. The role of G-protein in matrix-mediated motility of highly and poorly invasive melanoma cells. Int J Cancer 1991; 48:113-20.
324. Jain RK. Transport of molecules across tumor vasculature. Cancer Metastasis Rev 1987; 6:559-93.
325. Dvorak HF. Leaky tumor vessels: consequences for tumor stroma generation and for solid tumor therapy. Prog Clin Biol Res 1990; 354A:317-30.
326. Fidler IJ. The biology of human cancer metastasis. Acta Oncol 1991; 30:668-75.
327. Kaminski M and Auerbach R. Tumor cells are protected from NK-cell-mediated lysis by adhesion to endothelial cells. Int J Cancer 1988; 41:847-9.
328. Glinsky GV. Cell adhesion and metastasis: is the site specificity of cancer metastasis determined by leukocyte-endothelial cell recognition and adhesion? Crit Rev Oncol Hematol 1993; 14:229-77.
329. Walz DA. Thrombospondin as a mediator of cancer cell adhesion in metastasis. Cancer Metastasis Rev 1992; 11:313-24.
330. Bar-Shavit R, Ginzburg Y, Maoz M et al. The involvement of thrombin RGD in metastasis: characterization of a cryptic adhesive site. Israel J Med Sci 1995; 31:86-94.
331. Woodruff MF. The cytolytic and regulatory role of natural killer cells in experimental neoplasia. Biochim Biophys Acta 1986; 865:43-57.
332. Natali P, Nicotra Mr, Cavaliere R et al. Differential expression of intercellular adhesion molecule 1 in primary and metastatic melanoma lesions. Cancer Res 1990; 50:1271-8.
333. Harning R, Myers C, Merluzzi VJ. Monoclonal antibodies to lymphocyte function-associated antigen-1 inhibit invasion of human lymphoma and metastasis of murine lymphoma. Clin Exp Metastasis 1993; 11:337-42.
334. Crissman JD, Hatfield J, Schaldenbrand M et al. Arrest and extravasation of B16 amelanotic melanoma in murine lungs. A light and electron microscopic study. Lab Invest 1985; 53:470-8.
335. Starkey JR, Liggitt HD, Jones W et al. Influence of migratory blood cells on the attachment of tumor cells to vascular endothelium. Int J Cancer 1984; 34:535-43.
336. Orr FW and Warner DJ. Effects of neutrophil-medited pulmonary endothelial injury on the localization and metastasis of circulating Walker carcinosarcoma cells. Invasion Metastasis 1987; 7:183-96.

337. Auerbach R, Lu WC, Pardon E et al. Specificity of adhesion between murine tumor cells and capillary endothelium: an in vitro correlate of preferential metastasis in vivo. Cancer Res 1987; 47:1492-6.
338. Radinsky R, Fidler IJ. Regulation of tumor cell growth at organ-specific metastasis. In Vivo 1992; 6:325-31.
339. Rusciano D, Burger MM. Why do cancer cells metastasize into particular organs? Bioessays 1992; 14:185-94.
340. Patarroyo M, Lindbom L, Lundberg C. Leukocyte adhesion: molecular basis and relevance in inflammation. Advances in Exp Med Biol 1991; 314:1-187.
341. Bevilacqua MP. Endothelial-leukocyte adhesion molecules. Ann Rev Immunol 1993; 11:767-804.
342. Harning R, Myers C, Merluzzi VJ. Monoclonal antibodies to lymphocyte function-associated antigen-1 inhibit invasion of human lymphoma and metastasis of murine lymphoma. Clin Exp Metastasis 1993; 11:337-42.
343. Woynarowska B, Skrincosky DM, Haag A et al. Inhibition of lectin-mediated ovarian tumor cell adhesion by sugar analogs. J Biol Chem 1994; 269:22797-803.
344. Nakamori S, Kameyama M, Imoka S et al. Increased expression of sialyl Lewis^X antigen correlates with poor survival in patients with colorectal carcinoma: clinicopathological and immunohistochemical study. Cancer Res 1993; 53:3632-7.
345. Irimura T, Nakamori S, Matsushita Y et al. Colorectal cancer metastasis determined by carbohydrate-mediated cell adhesion: role of sialyl-Le^X antigens. Sem Cancer Biol 1993; 4:319-24.
346. Sawada R, Tsuboi S, Fukuda M. Differential E-selectin-dependent adhesion efficiency in sublines of a human colon cancer exhibiting distinct metastatic potentials. J Biol Chem 1994; 269:1425-31.
347. Iwai K, Ishikura H, Kaji M et al. Importance of E-selectin (ELAM-1) and sialyl Lewis^a in the adhesion of pancreatic carcinoma cells to activated endothelium. Int J Cancer 1993; 54:972-7.
348. Bresalier RS. Adhesion molecules and gastrointestinal malignancies. Gastroenterol 1994; 106:1378-83.
349. Jorgensen T, Berner A, Kaalhus O et al. Up-regulation of the oligosaccharide sialyl Lewis^X: a new prognostic parmeter in metastatic prostate cancer. Cancer Res 1995; 55:1817-9.
350. Groves RW, Allen MH, Ross EL et al. Expression of selectin ligands by cutaneous squamous cell carcinoma. Am J Pathol 1993; 143:1220-5.
351. Tozeren A, Kleinman HK, Grant DS et al. E-selectin-mediated dynamic interactions of breast-and colon-cancer cells with endothelial-cell monolayers. Int J Cancer 1995; 60:426-3l.
352. Brodt P. Characterization of two highly metastatic variants of the Lewis lung carcinoma with different organ specificities. Cancer Res 1986:46:2442-8.

353. Bani MR, Garofalo A, Scanziani E et al. Effect of interleukin-1-beta on metastasis formation in differnt tumor systems. J Natl Cancer Inst 1991; 83:119-23.
354. Orosz P, Kruger A, Hubbe M et al. Promotion of experimental liver metastasis by tumor necrosis factor. Int J Cancer 1995; 60:867-71.
355. Mannel DN, Orosz P, Hafner M et al. Mechanisms involved in metastasis enhanced by inflammatory mediators. Circul Shock 1994; 44:9-13.
356. Giavazzi R, Garofalo A, Bani MR et al. Interleukin 1-induced augmentation of experimental metastases from a human melanoma in nude mice. Cancer Res 1990; 50:4771-5.
357. Elices MJ, Osborn K, Takada Y et al. VCAM-1 on activated endothelium interacts with the leukocyte integrin VLA-4 at a site distinct from the VLA-4/fibronectin binding site. Cell 1990; 60:577-84.
358. Andrew DP, Berlin C, Honda S et al. Distinct but overlapping epitopes are involved in $\alpha_4\beta_7$-mediated adhesion to vascular cell adhesion molecule-1, mucosal addressin-1, fibronectin, and lymphocyte aggregation. J Immunol 1994; 153:3847-61.
359. Paavonen T, Tiisala S, Majuri ML et al. In vivo evidence of the role of $\alpha_4\beta_1$-VCAM-1 interaction in sarcoma, but not in carcinoma extravasation. Int J Cancer 1994; 58:298-302.
360. Hart IR, Birch M, Marshall JF. Cell adhesion receptor expression during melanoma progression and metastasis. Cancer Metastasis Rev 1991; 10:115-28.
361. Mould AP, Askari JA, Craig SE et al. Integrin $\alpha_4\beta_1$-medited melanoma cell adhesion and migration on vascular cell adhesion molecule-1 (VCAM-1) and the alternatively spliced IIICS region of fibronectin. J Biol Chem 1994; 269:27224-30.
362. Garofalo A, Chirivi RG, Foglieni C et al. Involvement of the very late antigen 4 integrin on melanoma in interleukin 1-augmented experimentl metastases. Cancer Res 1995; 55:414-9.
363. Zhu DZ, Cheng CF, Pauli BU. Mediation of lung metastasis of murine melanomas by a lung-specific endothelial cell adhesion molecule. Proc Natl Acad Sci USA 1991; 88:9568-72.
364. Zhu D, Cheng CF, Pauli BU. Blocking of lung endothelial cell adhesion molecule-1 (Lu-ECAM-1) inhibits murine melanoma lung metastasis. J Clin Invest 1992; 89:1718-24.
365. Zhu D, Pauli BU. Correlation between the lung distribution patterns of Lu-ECAM-1 and melanoma experimental metastases. Int J Cancer 1993; 53:628-33.
366. Cheresh DA. Human endothelial cells synthesize and express an Arg-Gly-Asp-directed adhesion receptor involved in attachment to fibrinogen and von Willebrand factor. Proc Natl Acad Sci USA 1987; 84:6471-5.

367. Cheresh DA. Integrins in thrombosis, wound healing and cancer. Biochem Soc Trans 1991; 19:835-8.
368. Felding-Habermann B, Cheresh DA. Vitronectin and its receptors. Curr Opin Cell Biol 1993; 5:864-8.
369. Lafrenie RM, Podor TJ, Buchanan MR et al. Up-regulated biosynthesis and expression of endothelial cell vitronectin receptor enhances cancer cell adhesion. Cancer Res 1992; 52:2202-8.
370. Lafrenie RM, Gallo S, Podor TJ et al. The relative roles of vitronectin receptor, E-selectin and $\alpha_4\beta_1$ in cancer cell adhesion to interleukin-1-treated endothelial cells. J Clin Invest 1992; 90:1406-13.
371. Humphries MJ, Olden K, Yamada KM. A synthetic peptide from fibronectin inhibits experimental metastasis of murine melanoma cells. Science 1986; 233:467-70.
372. Iwamoto Y, Robey FA, Graf J et al. YIGSR, a synthetic laminin pentapeptide, inhibits experimentl metastasis formation. 1988; 239:245.
373. Vollmers HP, Imhof BA, Braun S et al. Monoclonal antibodies which prevent experimental lung metastases. Interference with adhesion of tumor cells to laminin. FEBS Lett 1984; 172:17-20.
374. Sweeney TM, Kibbey MC, Zain M et al. Basement membrane and the SIKVAV laminin-derived peptide promote tumor growth and metastases. Cancer Metastasis Rev 1991; 10:245-54.
375. Taraboletti G, Belotti D, Giavazzi R et al. Enhancement of metastatic potential of murine and human melanoma cells by laminin receptor peptide G: attachment of cancer cells to subendothelial matrix as a pathway for hematogenous metastasis. J Natl Cancer Inst 1993; 85:235-40.
376. Hall CL, Yang B, Yang XW et al. Overexpression of the hyaluronan receptor RHAMM is transforming and is also required for H-ras transformation. Cell 1995; 82:19-26.
377. Hynes RO. The impact of molecular biology on models for cell adhesion. Bioessays 1994; 16:663-9.
378. Khoka R. Suppression of the tumorigenic and metastatic abilities of murine B16-F10 melanoma cells in vivo by the overexpression of the tissue inhibitor of the metalloproteinases-1. J Natl Cancer Inst 1991; 86:299-304.
379. Marchetti D, Menter D, Jin L et al. Nerve growth factor effects on human and mouse melanoma cell invasion and heparanase production. Int J Cancer 1993; 55:692-9.
380. Menter DG, Herrmann JL, Nicolson GL. The role of trophic factors and autocrine/paracrine growth factors in brain metastasis. Clin Exp Metastasis 1995; 13:67-88.
381. Vlodavsky I, Eldor A, Haimovitz-Friedman A et al. Expression of heparanase by platelets and circulating cells of the immune system: possible involvement in diapedesis and extravasation. Invasion Metastasis 1992; 12:112-27.

382. Bitan M, Mohsen M, Levi E et al. Structural requirements for inhibition of melanoma lung colonization by heparanase inhibiting species of heparin. Israel J Med Sci 1995; 31:106-18.
383. Brodt P. Selection of a highly metastatic liver-colonizing subpopulation of Lewis lung carcinoma variant H-59 using murine hepatocyte monolayers. Clin Exp Metastasis 1989; 7:525-39.
384. La Riviere G, Klein Gebbinck JW, Driessens MH et al. Pertussis toxin inhibition of T-cell hybridoma invasion is reversed by manganese-induced activation of LFA-1. J Cell Sci 1994; 107:551-9.
385. Kemperman H, Wijnands Y, de Rijk D et al. The integrin $\alpha_6\beta_4$ on TA3/Ha mammary carcinoma cells is involved in adhesion to hepatocytes. Cancer Res 1993; 53:3611-7.
386. Kemperman H, Wijnands Y, Meijne AML et al. TA3/St, but not TA3/Ha, mammary carcinoma cell adhesion to hepatocytes is mediated by $\alpha_5\beta_1$ interacting with surface-associated fibronectin. Cell Adhesion Commun 1994; 2:45-58.
387. Long Li, Nip J, Brodt P. Paracrine growth stimulation by hepatocyte-derived insulin-like growth factor-1: a regulatory mechanism for carcinoma cells metastatic to the liver. Cancer Res 1994; 54:3732-7.
388. Long L, Rubin RE, Baserga R et al. Loss of the metastatic phenotype in murine carcinoma cells expressing an antisense RNA to the insulin-like growth factor receptor. Cancer Res 1995; 55:1006-9.
389. Meterissian SH, Toth CA, Steele G Jr et al. Kupffer cell/tumor cell interactions and hepatic metastasis in colorectal cancer. Cancer Letters 1994; 81:5-12.

INDEX

Page numbers in italics denote figures (f) or tables (t).